T0312694

Basics of Ecotoxicology

Basics of Ecotoxicology

Donald W. Sparling

CRC Press
Taylor & Francis Group
Boca Raton London New York

CRC Press is an imprint of the
Taylor & Francis Group, an **informa** business

MIX
Paper from
responsible sources
FSC® C014174
www.fsc.org

CRC Press
Taylor & Francis Group
6000 Broken Sound Parkway NW, Suite 300
Boca Raton, FL 33487-2742

© 2018 by Taylor & Francis Group, LLC
CRC Press is an imprint of Taylor & Francis Group, an Informa business

No claim to original U.S. Government works

Printed on acid-free paper

International Standard Book Number-13: 978-1-138-03171-5 (Hardback)

Library of Congress Cataloging-in-Publication Data

Names: Sparling, D. W. (Donald W.), author.
Title: Basics of ecotoxicology / Donald W. Sparling.
Description: Boca Raton : Taylor & Francis, CRC Press, 2017. | Includes bibliographical references and index.
Identifiers: LCCN 2017005194| ISBN 9781138031715 (hardback) |
ISBN 9781315158068 (ebook)
Subjects: LCSH: Environmental toxicology. | Pollutants.
Classification: LCC RA1226 .S67 2017 | DDC 615.9/02--dc23
LC record available at https://lccn.loc.gov/2017005194

Visit the Taylor & Francis Web site at
http://www.taylorandfrancis.com

and the CRC Press Web site at
http://www.crcpress.com

Printed and bound in the United States of America by Sheridan

Contents

Preface...xi
Acknowledgments..xiii
Author xv

SECTION I   An Introduction to Ecotoxicology

Chapter 1   What Is Ecotoxicology? 3

    1.1   Introduction—What Do We Mean By 'Ecotoxicology'? 3
    1.2   How Did the Science of Ecotoxicology
           Come About? 4
    1.3   What Does It Take to Be an Ecotoxicologist? 7
    1.4   Objectives of This Book 7
    1.5   Chapter Summary 8
    1.6   Self-Test 8
    References 9

Chapter 2   Basic Chemistry for Ecotoxicologists 11

    2.1   Introduction 11
    2.2   Elements and Periodic Chart 11
           2.2.1   Elements 11
           2.2.2   Periodic Chart of Elements.............................12
    2.3   Chemical Reactions 14
    2.4   Chemical Groups of Greatest Interest to Ecotoxicologists......15
           2.4.1   Metals 15
           2.4.2   Organic Molecules 16
    2.5   Comments on the Fate and Transport of Chemical
           Contaminants in the Environment 18
           2.5.1   Metals 18
           2.5.2   Radioactive Elements 18
           2.5.3   Organic Molecules 18
    2.6   Chemicals of Major Interest 21
           2.6.1   Polychlorinated Biphenyls and Similar Chemicals 21
           2.6.2   Polycyclic Aromatic Hydrocarbons.....................21
           2.6.3   Organochlorine Pesticides.............................21
           2.6.4   Currently Used Pesticides 22
           2.6.5   Metals and Metalloids 22
           2.6.6   Other Chemicals...22
    2.7   Units of Measurement 22

2.8   Chapter Summary 23
2.9   Self-Test..24
References 25

Chapter 3   Some Ways Contaminants Affect Plants and Animals 27

3.1   Introduction 27
3.2   Generalized Effects of Contaminants 27
3.3   Physical or Mechanical Blockage.......................................28
3.4   Malformations 29
3.5   Interference with Enzyme Activity 31
       3.5.1   Chelation 31
       3.5.2   Receptor Binding..31
       3.5.3   Neurotransmitter Interference 32
       3.5.4   Effects of Contaminants on Plants 33
3.6   Impingements on the Immune System 33
3.7   Contaminant-Induced Endocrine Disruption.......................34
3.8   Genotoxicity 35
3.9   Cytochrome P450 and Other Metabolic Systems...................36
3.10   So How Do Ecotoxicologists Study These Effects?...............37
3.11   Chapter Summary 40
3.12   Self-Test..41
References 42

SECTION II   Major Groups of Contaminants: Where They Come from, What They Can Do

Chapter 4   Metals 47

4.1   Introduction 47
4.2   Sources of Metals in the Environment.................................47
4.3   Factors Affecting the Behavior of Metals............................49
4.4   Biological Effects of Metals 50
4.5   Characteristics of Lead and Mercury 52
       4.5.1   Characteristics of Lead..52
       4.5.2   General Characteristics of Mercury 56
4.6   Chapter Summary 61
4.7   Self-Test..61
References 62

Chapter 5   Current Use Pesticides...65

5.1   Introduction 65
5.2   What Is a Pesticide?..66
       5.2.1   Pesticide Use Is Controversial.................................66

5.3   Economics of Current Use Pesticides 67
5.4   Types of Pesticides 68
        5.4.1   Carbamates 69
        5.4.2   Organophosphates 70
        5.4.3   Pyrethroids 71
        5.4.4   Phosphonoglycine 74
        5.4.5   Triazines 76
        5.4.6   Inorganics, Metals, and Biologics 77
5.5   Chapter Summary 78
5.6   Self-Test 79
References 80

Chapter 6   Halogenated Organic Contaminants 83

6.1   Introduction 83
6.2   Polychlorinated Biphenyls 85
        6.2.1   Chemistry 85
        6.2.2   Persistence 87
        6.2.3   Breakdown of PCBs 87
        6.2.4   Concentrations of PCBs in Some Animal Tissues 88
        6.2.5   Biological Effects of PCBs 90
6.3   Ecotoxicity of Dioxins, Furans, and Dioxin-Like
        Compounds 90
        6.3.1   General Mechanisms of Toxicity 91
6.4   Polybrominated Diphenyl Ethers and Polybrominated
        Biphenyls 93
6.5   Polyfluorinated Organic Compounds 94
6.6   Chapter Summary 96
6.7   Self-Test 96
References 98

Chapter 7   Other Major Organic Contaminants 101

7.1   Introduction 101
7.2   Polycyclic Aromatic Hydrocarbons 101
        7.2.1   Chemical Characteristics of PAHs 102
        7.2.2   Sources and Uses of PAHs 103
        7.2.3   Persistence 103
        7.2.4   Environmental Concentrations 104
        7.2.5   Some Examples of Biological Concentrations 105
        7.2.6   Biological Effects of PAHs 105
7.3   Oil Spills and PAH 108
7.4   Organochlorine Pesticides 109
        7.4.1   Sources and Use 111
        7.4.2   General Chemical Characteristics of OCPs 111
        7.4.3   Structure 111
        7.4.4   Persistence 113

            7.4.5   Examples of OCP Concentrations in
                    Environmental Sources...114
            7.4.6   Concentrations of OCPs in Animals........................115
            7.4.7   Biological Effects of Organochlorine Pesticides......117
    7.5   Chapter Summary...119
    7.6   Self-Test..120
    References 121

Chapter 8   Contaminants of Increasing Concern 123
    8.1   Introduction 123
    8.2   Plastics...123
    8.3   Pharmaceuticals...128
    8.4   Nanoparticles...129
    8.5   Acid Deposition...132
    8.6   Chapter Summary...136
    8.7   Self-Test..137
    References 138

SECTION III   Higher Level Effects, Analysis of Risk, and Regulation of Chemicals

Chapter 9   Studying the Effects of Contaminants on Populations.....................143
    9.1   Introduction 143
    9.2   How Might Contaminants Influence the Characteristics
            of Populations?..145
            9.2.1   Density Dependence and Independence...................145
            9.2.2   Abundance...146
            9.2.3   Sex Ratios...148
            9.2.4   Age Structure..150
    9.3   Contaminants and Life Table Analyses 153
    9.4   Chapter Summary...156
    9.5   Self-Test..157
    References 158

Chapter 10 How Contaminants Can Affect Community and Ecosystem
            Dynamics...161
    10.1   Introduction 161
    10.2   Aspects of Community Ecology..161
            10.2.1 Species Richness, Diversity, and Abundance 161
            10.2.2 Food Chains and Webs 163
            10.2.3 Symbiotic Relationships...165
            10.2.4 Ecological Succession 166
            10.2.5 Community Sensitivity...168

10.3   Contaminants and Ecosystems 168
10.4   Chapter Summary 171
10.5   Self-Test 172
References 173

Chapter 11 Risk Assessment 175

11.1   Introduction 175
11.2   Brief Overview of the History of Risk Assessment 176
11.3   What Is an Ecological Risk Assessment Like? 176
11.4   Assessing Risk to Organisms 178
11.5   Uncertainty in Risk Assessments 181
11.6   Vulnerability Analysis 182
11.7   Risk Management 183
11.8   Chapter Summary 185
11.9   Self-Test 185
References 187

Chapter 12 Domestic and Global Regulation of Environmentally Important
Chemicals 189

12.1   Introduction 189
12.2   International Authorities 189
        12.2.1   The United Nations 189
        12.2.2   Organisation for Economic Co-Operation and
                 Development 191
        12.2.3   European Union 192
12.3   National Regulation of Contaminants 192
        12.3.1   State Department 193
        12.3.2   Department of Defense 194
        12.3.3   Department of Agriculture 194
                 12.3.3.1   U.S. Forest Service 194
                 12.3.3.2   Natural Resources Conservation
                            Service 194
        12.3.4   Department of the Interior 195
                 12.3.4.1   Fish and Wildlife Service 195
                 12.3.4.2   Bureau of Land Management 195
                 12.3.4.3   U.S. Geological Survey 196
                 12.3.4.4   Bureau of Safety and Environmental
                            Enforcement 196
        12.3.5   Department of Commerce 196
                 12.3.5.1   National Oceanic and Atmospheric
                            Administration 196
                 12.3.5.2   U.S. Coast Guard 197
        12.3.6   U.S. Environmental Protection Agency 197
                 12.3.6.1   Federal Food, Drug, and Cosmetic Act
                            (FFDCA, 1938) 198

12.3.6.2 Clean Air Act (1970)................................198
12.3.6.3 Clean Water Act (1972) 198
12.3.6.4 Resource Conservation and Recovery
          Act (RCRA, 1976)....................................199
12.3.6.5 Toxic Substances Control Act
          (TSCA, 1976) 199
12.3.6.6 Comprehensive Environmental
          Response, Compensation, and Liability
          Act (CERCLA, 1980) 200
12.3.6.7 Federal Insecticide, Fungicide, and
          Rodenticide Act (FIFRA, 1996) 201
12.4 Regulation at the State and Municipal Levels 201
12.5 Chapter Summary 202
12.6 Self-Test...202
References 203

Chapter 13 Future Perspectives and Concluding Remarks 205

13.1 Introduction 205
13.2 Does Looking Backward Tell Us Anything about Where
      the Science is Headed? 205
13.3 What Is the Current Status of Ecotoxicology? 207
13.4 Where Should the Science Head?...207
      13.4.1 Conduct Studies on the Huge Inventory Under
              the Toxic Substances Control Act (TSCA)...............208
      13.4.2 Effects of Multiple Contaminants on Organisms 208
      13.4.3 Increased Consideration by Regulatory Agencies
              for Relevant Species 208
      13.4.4 Alternative Methods 208
      13.4.5 Develop Realistic Scenarios 209
      13.4.6 Increase Study of Higher Level Effects....................209
      13.4.7 Gain More Information on Nanoparticles and
              Their Effects 209
      13.4.8 Encourage Advancements in Environmental
              Chemistry and Risk Assessment 209
References 210

Glossary 211

Index...215

Preface

To a large extent, this book started 20 years ago. I was a wildlife research biologist working for the U.S. Fish and Wildlife Service studying endangered forest birds in Hawaii when I received a transfer back to the home research center in Maryland. At that time, I was placed in the Environmental Contaminants Branch at Patuxent Wildlife Research Center because contaminant studies were of great interest to the U.S. Fish and Wildlife Service. To make a long story short, ultimately research within the U.S. Department of the Interior was moved around from one agency to another, finally landing within the U.S. Geological Survey where experimental research was not as appreciated as it was in the U.S. Fish and Wildlife Service. Over the years, I worked with a variety of animals, including mallards, northern bobwhite, and several species of frogs, salamanders, mourning doves, and turtles. These animals were variously exposed to multiple pesticides, white phosphorus, polycyclic aromatic hydrocarbons, contaminated sediments, lead, aluminum, and mercury.

The field of ecotoxicology is broad and includes ecologists, wildlife physiologists, environmental chemists, statisticians, modelers, and others. The whole time I worked in ecotoxicology my specialty area was wildlife toxicology or the effects contaminants had on wildlife individuals, populations, and communities. While I was also interested in how the chemicals moved through the environment, from water to sediments to organisms, I was never a research chemist and left those analyses up to cooperators. My general approach has been that contaminants are another source of stress to individuals and to wildlife populations, much like predation, disease, inter- and intraspecific competition, and resource allocation and that their effects could be studied in similar ways but using different methods.

After many years working within the U.S. Department of the Interior, I took a faculty position at my old alma mater in the Zoology Department and continued research in wildlife toxicology and added teaching to my resume. I also wrote and published a significantly sized tome in ecotoxicology.

It was during my efforts to educate college students and in preparing that book that I came to realize that there was a gaping hole in the offerings of ecotoxicology-related books on the market. Many of the available books were lengthy, somewhat to very ponderous volumes that were deficient in simplicity and clarity. It was then that I resolved to attempt to publish a book that was written in clear, easily understandable English so that undergraduates and perhaps even advanced placement high-school seniors could understand the science of ecotoxicology.

This, then, is the goal of this current text. I hope that students without an extensive (dare I say "no"?) background in chemistry can read and understand what has been written here. I do assume that students have had at least a college-level course in biology and hopefully an ecology or conservation course, but I hope I have written a primer on chemistry that, with the instructor's lectures, will provide sufficient understanding of that discipline's core elements even with no formal training in chemistry. I will leave it to the student to determine if I have been successful.

In this book key words are indicated in bold type. There are too many key words to include in a glossary and some of these words are easily recognized to not all are included in the glossary. Phrases of particular interest are often surrounded by single quotations whereas actual quotes are defined by double quotation marks.

This book consists of three sections. Section I consists of some basic information, including the definition and inclusiveness of ecotoxicology, the chemistry primer, and another primer on the various ways contaminants can affect wildlife. Section II covers the primary classes of contaminants, including inorganic and organic molecules. Each chapter discusses something about the history of the class, what it was or is used for, how the chemicals behave in the environment, and the expected effects caused by contaminants in this class. Section III covers higher-level discussions on how contaminants affect populations, communities, and ecosystems; methods used to determine the real and predictable harm chemicals can inflict on organisms and their habitats; and how contaminants are regulated at the local, national, and international levels. I hope that you find this book enjoyable to read as well as informative.

Acknowledgments

Behind every book there are many people who are involved in its writing, even if they do not know it. Such is the case with this one. I first thank Dr. Elwood Hill, who was my early mentor when I began the study of ecotoxicology in the 1980s. Woody had a lot of patience for a younger PhD student who did not know much about environmental contaminants at all. He taught me the basics of the science and was often there when I needed support. I also thank a group of biologists at Patuxent Wildlife Research Center, including Drs. Russell Hall, David Hoffman, Barnett Rattner, and Hank Pattee for their guidance. Many more scientists, technicians, and students helped form my professionalism and challenged me to be my best. Drs. Greg Linder and Christine Bishop along with Sherri Krest were coeditors with me on four other books that finally led to this one. Special thanks to my wife, Paulette, and family for their support and patience.

Author

Dr. Donald Sparling's career has been a mixture of college-level teaching and government service. After receiving his PhD in biology, he taught wildlife management courses at Southern Illinois University and biology at Ball State University. In 1982, he started a career in the U.S. Department of the Interior, first as a statistician and then as a research wildlife biologist. There he conducted research on the effects of contaminants on wildlife, with publications on acid deposition, munitions, pesticides, metals, polycyclic aromatic hydrocarbons, polychlorinated biphenyls, and polybrominated diphenyl ethers. The animal models he and his students have used include amphibians, turtles, mallards, quail, doves, invertebrates, and others. He left the government in 2004 to resume teaching various courses including ecotoxicology and to continue research at Southern Illinois University, Carbondale. Although he retired from that position in 2015, he continues to teach online courses at Holy Apostles College and Seminary and write books in the ecological sciences. Dr. Sparling has more than 100 scientific publications and 6 books to his credit.

Section I

An Introduction to Ecotoxicology

1 What Is Ecotoxicology?

1.1 INTRODUCTION—WHAT DO WE MEAN BY 'ECOTOXICOLOGY'?

Since we are going to spend some time studying the subject, we probably should discuss what the word '**ecotoxicology**' means and what its study includes. If we take the word apart we find 'eco' and 'toxicology.' You are probably familiar with the prefix 'eco' as in eco-friendly, eco-systems, and eco-logy. These uses of 'eco' all pertain to the science of understanding the environment, otherwise known as ecology. 'Toxicology' is the study of poisons. The two terms were first brought together in a lecture provided by René Truhaut in 1969 to denote a natural extension of ecology and toxicology that included the effects of chemical pollutants or **contaminants** on any aspect of the environment (Truhaut, 1977). A concern about contaminants existed in earnest in the Western developed countries in the mid-1960s, but there wasn't a comprehensive term for the newly developing science until then.

In its early years, ecotoxicology typically included two things: (1) where chemical contaminants were located as well as how much was present and (2) a basic understanding of how toxic individual chemicals can be to single species of wildlife and plants, with an emphasis on short-term or **acute** studies. The greatest emphasis has always been on human-caused or **anthropogenic** sources of environmental chemicals, although many instances of natural pollution have also been studied.

Over the years the chemical aspect of ecotoxicology has expanded to include the life cycle of contaminants—how they get into the environment, what changes occur when they are there, and how natural processes or human activities can reduce or eliminate these contaminants. From the toxicological side there has been growing interest on longer-term or **chronic** studies that may occur over weeks, months, or a particular stage in an organism's life, such as the larval stage. More recently, ecotoxicologists are learning how interactions of chemicals may affect organisms and are finding some very interesting results. There is also a developing interest in modeling the potential effects of one or more chemicals on selected species without the need for controlled experimentation.

Some fundamental discoveries in ecotoxicology over the past 50 years include the following—we will spend more time with each of these and many more concepts in future chapters:

1. The half-life of chemicals in the environment ranges widely. A **half-life** is the amount of time it takes to degrade 50% of an original concentration of a contaminant. Following the dynamics of degradation, this means that if we had an initial concentration of a chemical in soil, for example, of 10 parts per million (ppm) and the chemical had a half-life of 20 years, in 20 years 50% of the chemical would be gone, leaving a concentration of 5 ppm, in 40 years there would be 2.5 ppm, in 60 years 1.25 ppm, and so on.

2. Environmental chemists have developed an understanding that some chemicals are extremely persistent and can last in the environment for decades. The poster child, so to speak, for these persistent chemicals is dichlorodiphenyltrichloroethane (DDT), which can have a half-life in soil of 20–30 years or more, depending on conditions (ATSDR, 2002). Extending a few half-lives, a single instance of DDT contamination of a few hundred parts per million could still be measured more than a hundred years later. In contrast, other contaminants can disappear in a few hours when exposed to sunlight. Half-lives depend on many factors, but where the contaminant is located is paramount—those in the atmosphere tend to disappear most rapidly, those in deep water sediments the most slowly.

3. Early on, biologists determined that the toxicity of a given chemical was dependent, in part, on the length of time organisms are exposed to the chemical. As mentioned, early studies focused on the concentration or amount of a contaminant that resulted in death, a **lethal dose**. In general, the shorter the duration, the higher the concentration a chemical must have to be lethal. At longer durations, it usually takes lower concentrations of a chemical to cause death in a substantial proportion of a population. In addition, other **sublethal effects** may occur in chronic exposures, such as reduced growth, developmental defects, or cancers; acute exposures may be too short to make these effects apparent.

4. Toxicity also varies across species. It is a generalization that aquatic organisms such as fish or aquatic invertebrates are more sensitive to chemicals than terrestrial species such as birds or mammals. A major factor affecting this is that aquatic organisms are surrounded by contaminated water and everything they encounter has burdens, while birds and mammals are usually exposed only through food or by drinking and hence do not experience the degree of exposure that fish do. Even within a related group of organisms, such as birds, toxic concentrations may vary by 10 or 20 times across species. For example, Story et al. (2011) found that fenitrothion, an organophosphorus pesticide (see Chapter 5 for more information on this type of pesticide), differed by 26% in two species of Australian rodents, and both were 10–14 times more sensitive than house mice (*Mus musculus*) and other rodents.

5. Now that toxic concentrations have been identified for many contaminants and species, attention is increasingly being turned toward the effects of contaminants on entire populations, communities, and even ecosystems. New research is also focused on the interactive effects of contaminants.

1.2   HOW DID THE SCIENCE OF ECOTOXICOLOGY COME ABOUT?

Both ecology and toxicology have long histories, although not always in the form of sciences that we might recognize today. Environmental concerns go back to antiquity, with the first writings credited to Aristotle (384–322 BC) and his student Theophrastus (371–287 BC), both of whom described interrelationships between animals and their environment as early as the fourth century BC. However, the interest remained only a part of the broader science of biology until 2300 years later when ecotoxicology was recognized as a separate discipline.

People have been poisoning each other for an even longer time, but toxicology as a formal study probably began with Paracelsus, a German-Swiss philosopher, physician, botanist, astrologer, alchemist, and general occultist whose real name was Philippus Aureolus Theophrastus Bombastus von Hohenheim (1493–1541). Among other things, he is credited with the saying that "the poison is in the dose." Bombastus showed that all things, even water, can be poisonous in sufficient quantities. Some substances require a large dose before being toxic, whereas others only require a small dose, perhaps only a few milliliters for a human. In very small amounts, even substances that are considered highly toxic do not produce any symptoms. Bombastus is known as the 'Father of Toxicology.'

The marriage of ecology and toxicology occurred in the early 1960s and gained a strong boost in popular interest through the publication of *Silent Spring* by Rachel Carson (Figure 1.1) in 1962. Carson wrote about the possible dangers of DDT and other pesticides and how we were poisoning our environment in many ways. There was an acute rise in interest for protecting our environment in the late 1960s, resulting in the first Earth Day in Washington, DC, on April 22, 1970 (Figure 1.2), that rocked

FIGURE 1.1   Rachel Carson, the author of *Silent Spring*, 1962. (Courtesy of U.S. Fish and Wildlife Service, Washington, DC.)

FIGURE 1.2 Huge crowds gathered to hear speeches on the Mall in Washington, DC, during the first Earth Day in 1970. (Courtesy of National Park Service, Washington, DC.)

the country and the Congress. If we look at a timeline of scientific publications going back several decades, we see a very sharp increase in publications, indicating a growing research interest in ecotoxicology, starting in the late 1960s and rising exponentially since then (Figure 1.3). This figure shows only those publications dealing with pesticides, but similar curves exist for other contaminants.

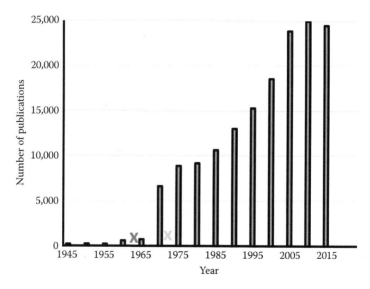

FIGURE 1.3 Annual number at 5-year intervals of pesticide-related publications in ecotoxicology. Publications on other contaminants are not included. The red X marks the beginning of an interest in the effects of contaminants in the environment and the green X is when *Silent Spring* was published. (Data from Web of Science, Washington, DC.)

1.3 WHAT DOES IT TAKE TO BE AN ECOTOXICOLOGIST?

Ecotoxicology is a rigorous discipline that has chemistry, physiology, and ecology as its main components. Chemistry is important in knowing the makeup and characteristics of the contaminants of interest, including their probable behavior in the environment. Physiology is useful in understanding how these chemicals may affect organisms at the subcellular, cellular, organ, or overall homeostasis levels. Ecology puts these two elements into real-life environmental contexts and asks how they might affect individual survival and reproduction as well as asking bigger questions of populations, communities, and ecosystems.

Statistics and modeling are also often used in ecotoxicological studies. Statistics are essential in designing studies and research experiments and in analyzing the data that are obtained. Modeling and statistics are very useful in evaluating the potential threats of contaminated areas to living organisms in a discipline called **risk assessment**.

In the ever-evolving world of ecotoxicology very few, if any, scientists can be experts in all of these specialties. Subspecialty fields exist in that some chemists may focus on a particular group of chemicals such as pesticides, polyhalogenated compounds (see Chapters 6 and 7), or metals. 'Effects people' may specialize in certain organisms such as plants, fish, aquatic invertebrates, amphibians, birds, or mammals.

Because of the complexity of the science, most practitioners have advanced or graduate degrees in their areas of specialty. With a bachelor's degree one might look for jobs as technicians or field assistants. A master's degree and work experience may be required for being a leader in investigations. Overseeing scientific investigations and research experiments often requires a PhD in one of the specialty areas listed above. A few universities offer programs specifically in ecotoxicology and a far larger number of support-related courses.

1.4 OBJECTIVES OF THIS BOOK

In writing this book, I hope to offer an introduction or overview of ecotoxicology for the science major who may not necessarily be headed toward a career in ecotoxicology or even graduate school. I hope that I have written a book that avoids much of the jargon and specific language of a specialty in ecotoxicology so that this book can be easily read and enjoyed by undergraduates. In this way, I hope that more students can obtain an understanding of ecotoxicology and through this develop a better appreciation for our environment and the complex interactions between it and contamination, more familiarly called pollution.

We will cover a variety of topics, including a brief introduction or refresher of chemistry and an examination of some of the ways that contaminants can affect organisms. These are found in Section I of this text. Section II deals with the major groups of contaminants including metals, a whole variety of organic contaminants, plastics, and even radioactive compounds. The final section includes several areas of current and developing concern, including contaminant-caused effects in higher levels of ecological organization. It also includes a section on how contaminated sites are evaluated and on agencies that regulate pollution from the global to the city levels. The glossary at the end of this book assists students with difficult or new terminology.

This book is not intended to provide a thorough and detailed explanation of ecotoxicology and its many subdisciplines, as might be desired by advanced undergraduates or graduate students. I have recently published one such tome (Sparling, 2016) as have others (see, e.g., Walker et al., 2012 or Newman, 2014). I hope, however, that you enjoy this shorter excursion into this fascinating discipline.

1.5   CHAPTER SUMMARY

1. The term 'ecotoxicology' was coined in the 1970s to refer to the study of the occurrence and effects of chemical contaminants in the environment.
2. Ecotoxicology involves both chemistry and ecology.
3. Initial endeavors in environmental chemistry focused on how much of a contaminant was present in an area. Later developments included the fate and transport of these chemicals, that is, how they degrade and get transferred from one part of the environment to another.
4. Early studies on the effects of contaminants on organisms explored how toxic chemicals were classified by determining their lethal concentrations in acute exposures. As the science developed, the effects of chronic exposures of one or more chemicals simultaneously became more numerous, as did studies of effects on populations, communities, and ecosystems.
5. Ecotoxicology requires expertise in chemistry, physiology, ecology, and other sciences such as statistics and computer modeling. Most scientists specialize in an area such as environmental chemistry of a specific group of compounds or the effects on a class of organisms. Graduate degrees in these fields are often required to pursue a career in this science.

1.6   SELF-TEST

1. What words form the root of the term 'ecotoxicology' and what do they mean?
2. Practitioners of ecotoxicology should have at least working knowledge in what disciplines?
   a.   Chemistry
   b.   Physiology
   c.   Ecology
   d.   All of the above
3. Studies on long-term (weeks or months) exposure to contaminants are called
   a.   Population studies
   b.   Chronic exposures
   c.   Lethal exposures
   d.   Acute exposures

4. The half-life of chlorpyrifos, a commonly used insecticide, is around 80 days in soil. If we started with a concentration of 100 ppm of the compound in soil, how much would be left after 320 days, assuming that the only loss was due to degradation and that no more insecticide was added in that time?
5. What book was extremely important in first making the American public aware of the potential dangers of contaminants in the environment? Who was its author?
6. Let us do a bit of mathematical computation to get an even better idea of the rapid growth of ecotoxicology as a science. In 1970 approximately 700 scientific papers were published on pesticides (see Figure 1.3). In 2010 around 25,000 papers were published. What was the rate of growth in scientific papers over those years?
7. Define in your own terms what is meant by a contaminant? Are chemicals the only form of contaminant in the environment or can there be other types of contaminants?
8. Who is considered to be the 'Father of Modern Toxicology'?
   a. Aristotle
   b. Truhaut
   c. Bombastus
   d. Carson
9. True or False. Chemical contaminants in the environment at low concentrations are more likely to produce sublethal effects than killing organisms outright.
10. True or False. The lethal dose of the pesticide chlorpyrifos for northern bobwhite quail can be used reliably for all species of birds.

REFERENCES

Agency for Toxic Substances and Disease Registry (ATSDR). 2002. *Toxicological profile for DDT, DDE, and DDD.* Atlanta, GA: U.S. Department of Health and Human Services, Public Health Service.

Carson, R. 1962. *Silent Spring.* New York: Houghton Mifflin.

Newman, M.C. 2014. *Fundamentals of Ecotoxicology: The Science of Pollution,* 4th Edition. Boca Raton, FL: CRC Press.

Sparling, D.W. 2016. *Ecotoxicology Essentials: Environmental Contaminants and Their Effects on Plants and Animals.* Cambridge, U.K.: Academic Press.

Story, P., Hooper, M.J., Astheimer, L.B., Buttemer, W.A. 2011. Acute oral toxicity of the organophosphorus pesticide fenitrothion to fat-tailed and stripe-faced dunnarts and its relevance for pesticide risk assessments in Australia. *Environ Toxicol Chem* 30: 1163–1169.

Truhaut, R. 1977. Eco-toxicology—Objectives, principles and perspectives. *Ecotox Environ Safe* 1: 151–173.

Walker, C.H., Sibley, R.M., Hopkins, S.P., Peakall, D.B. 2012. *Principles of Ecotoxicology,* 4th Edition. Boca Raton, FL: CRC Press.

2 Basic Chemistry for Ecotoxicologists

2.1 INTRODUCTION

Because ecotoxicology deals with chemicals in the environment and how they affect organisms, a certain level of understanding of chemistry is needed to fully appreciate the science. However, many of the readers of this book may not have had a lot of (or any) formal training in chemistry, and others who have had some classes may benefit from a review of this field. In this chapter, I hope to provide enough chemistry so that the rest of this book makes sense, and I hope I don't bore those who have had chemistry courses. We will very briefly cover the elements and the Periodic Chart. Then we will discuss how elements combine to form molecules and enter a cursory examination of the two major classes of molecules in ecotoxicology—metals and organics. Finally, we will review some concepts dealing with the fate and transport of chemical contaminants in the environment. If you have had training in chemistry and feel comfortable with the topic, you might want to skip over the parts you know well and focus on those sections that you may benefit from some refreshing. Students who desire greater information on basic chemistry may wish to consult one of the many texts on chemistry (e.g., Moore, 2011; Silberberg and Amateis, 2014).

2.2 ELEMENTS AND PERIODIC CHART

2.2.1 ELEMENTS

All matter is composed of elements. Some examples of elements include familiar ones such as iron, manganese, calcium, and sodium; some less familiar ones are californium, promethium, or neodymium. Most elements were discovered many years, even centuries, ago, but several have been identified through experiments with nuclear reactors. So far, 118 elements have been identified. While most elements are extremely stable, some of the ones discovered during experiments in nuclear reactors have very short life spans. For instance, flerovium has a half-life of 30 seconds and copernicium's half-life is less than 1/1000 of a second (Los Alamos National Laboratory, 2016). A radioactive half-life is the amount of time for the nucleus of the atom to decay. Ninety elements occur in nature in appreciable amounts, and another 4–8 are in trace quantities as a result of radioactive decay.

Elements are identified by the number of protons in their nucleus, which is called the **atomic number**. Hydrogen, the smallest of atoms, has only one proton in its nucleus, whereas lead has 82 protons. Protons have a positive charge, electrons have a negative charge, and neutrons have no charge. A neutral atom (one that has no net charge) has the same number of electrons as protons. Protons and neutrons have an

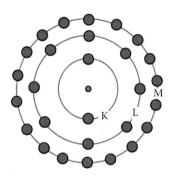

FIGURE 2.1   Valence shells in an atom contain the electrons around the nucleus.

atomic mass of 1 amu whereas electrons have no mass. The atomic mass of an element, therefore, is the total number of protons and neutrons in its nucleus.

While the nucleus of an atom is composed of protons and neutrons, electrons orbit the nucleus within **energy shells**. There are seven energy shells designated K through Q, each with an increasing number of subshells or orbital pathways that contain electrons (Figure 2.1). As the energy shell level increases from K to Q, the total number of electrons it can hold in the neutral state increases, with the lowest shell K holding only 2 electrons and the highest shell potentially holding 72 electrons, but no known shell has more than 32 electrons. While the exact location of an electron cannot be determined, they tend to travel in suborbits labeled s, p, d, and f. Electrons in the outermost orbits have higher average energy and travel farther from the nucleus than those in the inner orbits. Outermost electrons are important in determining how the atom reacts chemically and behaves as an electrical conductor. Because the pull of the atom's nucleus upon them is weaker and more easily broken than those that are closer to the nucleus, these electrons can be used in chemical reactions (see below). The outermost shell in an atom is called the **valence shell** and cannot hold more than eight electrons. There is a strong tendency among atoms to either fill this outer shell or transfer the electrons, whichever requires less energy.

2.2.2   PERIODIC CHART OF ELEMENTS

Due to the regularity of elements, we can systematically place them in order based on similar characteristics resulting in the **Periodic Chart of Elements** (Figure 2.2), with elements being placed into rows (periods) and columns (groups) based on their atomic number. Each group contains elements with the same number of electrons (in the neutral state) in the outermost shell. For instance, in the first column consisting of hydrogen (H) through francium (Fr); all contain one electron in the outermost or valence shell. This single electron is only loosely attracted by the protons in the nucleus and is readily donated so that the new outermost shell has a full complement of eight electrons. Elements in the 17th column or group, that is, fluorine (F) through astatine (As), are called **halogens** and have seven electrons in their outer shell, one short of filling that shell; these readily accept an electron. Elements in groups 1, 2, and 13 through 16 have two to six electrons in their outer shells. Those in groups 3

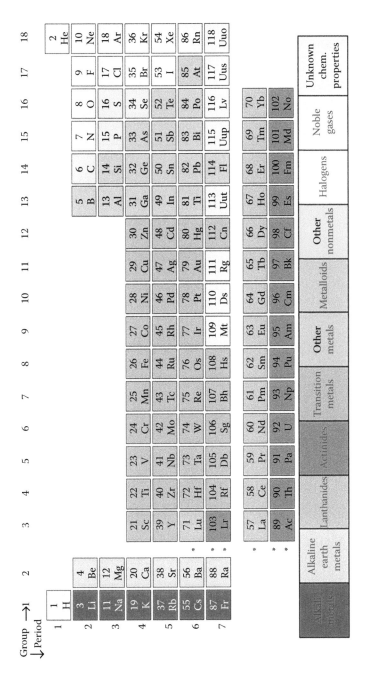

FIGURE 2.2 The Periodic Chart of elements. Elements are arranged in rows and columns based on the number and location of electrons. They are placed in classes based on their characteristics. (Courtesy of Coffield, C., Period table of elements with free printable, Multiplication.com, http://www. multiplication.com/our-blog/caycee-coffield/periodic-table-elements-class-project-free-printable, 2015.)

through 12 are called transition metals because they use electrons from more than one shell to form reactions. Elements on the far end including helium (He), neon (Ne), argon (Ar), krypton (Kr), xenon (Xe), and radon (Rn) have the full complement of eight electrons in the outermost subshell; they all occur in the gaseous state at standard temperature (25°C) and are very stable—hence, they are called **inert or noble gases**.

2.3 CHEMICAL REACTIONS

Elements can react with each other to form molecules. They do so by either (1) donating one or more electrons to the other element to fill the outermost shells, (2) accepting electrons from other elements, or (3) sharing electrons. Let's take a familiar example, that of table salt, which is composed of one atom of sodium (Na) and one atom of chlorine (Cl). Sodium is an alkali metal in the first group on the Periodic Chart, which means that it has a high propensity to donate an electron. Chlorine is a halogen in group 17, meaning that it has a high propensity to accept an electron to make its outermost shell filled. When Na and Cl are combined, sodium donates its electron to chlorine, which accepts it, and they make NaCl, or sodium chloride (Figure 2.3a).

Due to the readiness of gaining or losing electrons, molecules that involve either halogens or alkali metals tend to be very soluble in water. When dissolved in water, such molecules disassociate into their ionic forms. In the case of NaCl, this means that the molecule breaks apart into Na^+ and Cl^-, with the + (positive) and − (negative). For these reasons, the types of bonds formed between such elements are termed **ionic bonds** (Figure 2.3b).

Many elements such as calcium (Ca) have two or more electrons in their outermost shell and form stable molecules by donating both electrons. When combined with elements such as chlorine that accepts these electrons, a more complex molecule of $CaCl_2$ forms, with the 2 representing two atoms of chlorine for every calcium atom. This molecule is also water soluble and disassociates to Ca^{2+} and $2Cl^-$. Similarly, the oxygen in Figure 2.3b needs two electrons to fill its outermost shell and will combine with two hydrogens to do so.

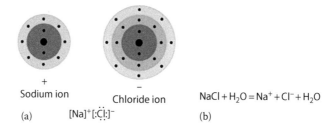

$$NaCl + H_2O = Na^+ + Cl^- + H_2O$$

Sodium ion   Chloride ion

(a) $[Na]^+ [:\ddot{C}l:]^-$ (b)

FIGURE 2.3 A chemical reaction illustrating the disassociation of NaCl in water. (a) In this reaction, the water molecule is unaffected. In (a) the sodium and chloride ions are shown; in (b) the reaction of salt in water resulting in the disassociation of sodium and chlorine is shown. The number 2 following the hydrogen (H) atom indicates that two hydrogens combine with one oxygen (O) to form water (b).

$$Ca = S$$

FIGURE 2.4 An example of a double bond. Calcium (Ca) has two electrons in its outer shell that it can donate, and sulfur (S) can accept two electrons.

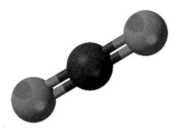

FIGURE 2.5 A ball-and-stick model for carbon dioxide illustrating the covalent bonds. In this model, carbon is shown in black and oxygen in red. (Courtesy of Wikimedia Commons, https://commons.wikimedia.org/wiki/File:Carbon_dioxide_structure.png.)

Elements with two electrons in their outermost shell can also react with those requiring two electrons to form a **double bond** (Figure 2.4). In this case, both electrons are donated. The double bonds tend to be stronger than single bonds, and such molecules tend to be less water soluble than those described above. Following the same train of thought, some molecules can have triple bonds, but that is as far as it goes; there are no naturally occurring quadruple-bonded molecules. However, a given element such as carbon can form four separate bonds with other elements.

By the time we reach the middle of the table, something different happens. Elements in Group 14 have four electrons in their outer shells. Thus, they can accept or donate equally. Instead of doing either of those, however, elements in this group have a high affinity for the electrons and are more likely to share them than give them to other atoms. When they share, electrons are sometimes closer to the nucleus of one atom and sometimes to the other. These types of bonds are called **covalent bonds**. For instance, carbon can react with two atoms of oxygen to form carbon dioxide (Figure 2.5). In this case the electrons that form the bonds between the atoms may be anywhere between the oxygens and carbon.

2.4 CHEMICAL GROUPS OF GREATEST INTEREST TO ECOTOXICOLOGISTS

There are two major groups of chemicals that most often concern ecotoxicologists: metals and organic molecules.

2.4.1 METALS

Metals occupy a majority of the naturally occurring elements, but they form different groups with somewhat different properties. The common properties among the metals are that most are **solid** at standard temperature, **ductile** (able to be drawn into thin wires), **malleable** (able to be hammered into thin sheets), **lustrous** (shiny), **good conductors of heat and electricity, and have high density and high melting point** (the temperature at which they become liquid). The glaring contrast to many of these characteristics is mercury, which is a liquid at standard temperature. The alkali metals (group 1) and alkaline earth metals (group 2) readily form ionic bonds with many other elements and have stable valence states of +1 or +2. The transitional metals are a bit more interesting. Because they can donate electrons from more than one subshell, they can take on several valence states. For example, chromium (Cr) can take on any valence state from 0 (solid chromium) to +6. Thus, metals are very flexible at forming reactions with other elements.

In ecotoxicology, a certain group of transitional metals known as **heavy metals** is of particular interest. As a broad definition, heavy metals are those that are denser than iron, but the ones of greatest concern include zinc (Zn), copper (Cu), chromium, cadmium (Cd), mercury (Hg), and lead (Pb). These metals are sufficiently prevalent in nature and in industrial uses and are moderately to highly toxic to living organisms to merit extra attention. We will spend a lot more space in Chapter 4 discussing the occurrence, use, and toxicity of these metals.

Another group that contains some elements of interest is the **metalloids**. These share some but not all of the characteristics of metals. Two elements in this group stand out—silicon (Si) and arsenic (As). Silicon, while not among the most toxic of elements, is the primary constituent of asbestos, which is well known for its relation to lung cancer. Arsenic, of course, has been widely used in many industrial products and is a well-known poison.

2.4.2 ORGANIC MOLECULES

Organic chemistry is a branch of chemistry that focuses on carbon and the elements it reacts with. Organic chemistry is very important to living organisms because carbon, oxygen, hydrogen, and nitrogen are the top four elements in living organisms. Carbon, second to oxygen in weight, makes up around 18% of humans by weight and forms the foundation for the majority of the molecules involved in life, including lipids, carbohydrates, proteins, and nucleic acids.

The simplest forms of organic molecules are **hydrocarbons** that are made up only of carbon and hydrogen. Examples of familiar hydrocarbons include methane gas (CH_4) and octane (C_8H_{18}, Figure 2.6). Hydrocarbons may occur as straight chains as alkanes with only single bonds between all the carbons (as in Figure 2.6b). Alkenes have at least one double bond between carbons and alkynes have at least one triple bond (Figure 2.6d and e). They may also occur as branched molecules. Figure 2.6c has the same number of carbon and hydrogen atoms as Figure 2.6b and has the same molecular weight. They are **isomers** of each other. There are also cycloalkanes that are circular in formation (Figure 2.7). As we will see in future chapters, cyclic organic molecules are very important to the formation of many contaminants.

While hydrocarbons, by definition, are formed only from carbon and hydrogen, as in benzo[a]pyrene in Figure 2.7, they can be the basis for many, many other molecules by combining with other elements. Chief among these additions for living

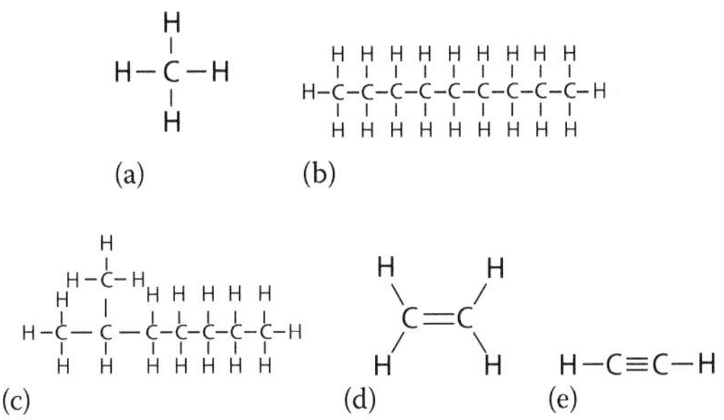

FIGURE 2.6 Examples of some hydrocarbons: (a) methane, (b) octane, (c) isooctane, (d) double-bonded ethylene, and (e) triple-bonded acetylene. Note that isooctane is a branched hydrocarbon and is an isomer of octane.

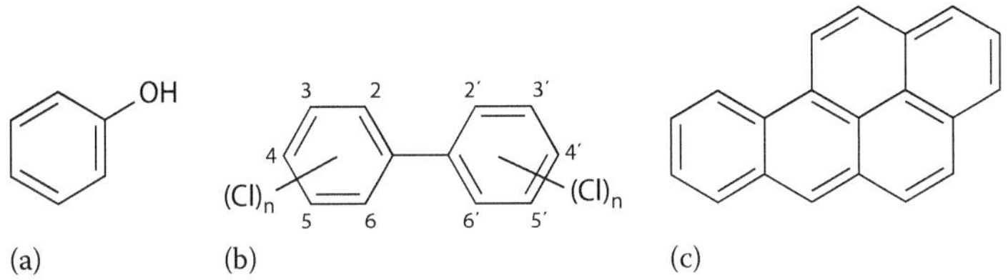

FIGURE 2.7 Examples of cyclo-organic molecules: (a) phenol, (b) polychlorinated biphenyl (PCB), and (c) benzo[a]pyrene. Each juncture represents a carbon atom and includes sufficient hydrogens to fill up the four places on the carbon atom. Phenol has a hydroxyl (OH) moiety attached. The PCB molecule contains chlorine and is referred to as a chlorinated hydrocarbon. The numbers represent specific carbon atoms within the PCB framework.

organisms is oxygen, with lipids and carbohydrates being entirely formed by these three elements. Other elements of consequence include nitrogen, sulfur, chlorine, and several others. All proteins and nucleic acids, for example, contain nitrogen.

Cyclic organic molecules can be composed only of hydrogens and carbons; these are referred to as saturated molecules. When other groups such as hydroxyls ($-OH^-$), amines ($-NH^{-2}$), or a host of others combine with the carbons, they 'bump' off hydrogens, and the resulting molecule is said to be **substituted**. One group of saturated cycloalkanes of interest to ecotoxicologists is polycyclic aromatic hydrocarbons (PAHs). When benzene, a saturated cycloalkane, is combined with an $-OH$ group (hydroxyl), it becomes **phenol** (Figure 2.7), and this forms the foundation for many other contaminants of interest.

A note concerning the diagrams of organic molecules is warranted. We will use structural formulas of molecules throughout this book. Full representation of these molecules would include all of the hydrogen atoms. However, by standard convention, it is understood that empty spots are filled with hydrogen atoms, thus resulting in the shorthand notation for the molecules in Figure 2.7.

2.5   COMMENTS ON THE FATE AND TRANSPORT OF CHEMICAL CONTAMINANTS IN THE ENVIRONMENT

2.5.1   METALS

Above, we mentioned about half-lives and having a half-life implies that a chemical can be broken down or degraded in the environment. Whereas metals are elements and cannot be degraded, they can be moved from one area of the environment to another. According to the Agency for Toxic Substances and Disease Registry (ATSDR, 2003), a government agency overseeing toxic substances, the principal reservoir for zinc, that is, where most of it is found in the environment, is the earth's crust. Through erosion, some of this may enter waterways such as streams or rivers. Because zinc is an essential nutrient, plants absorb it through their roots and animals ingest it by eating plants, other animals, or small amounts of dirt. Animals and plants excrete excess zinc, which results in a biological cycle for the metal.

An alternative route is for metals to eventually sink into deep ocean sediments and lie dormant for eons. Eventually, some of this metal may be returned to the surface through volcanoes or major uplifting such as mountain building.

Human actions can alter these natural cycles. Through mining, smelting, and industrial use, fine particles of metals may enter the atmosphere. They generally don't last long in this compartment but fall back to earth either in water or land. Anthropogenic (human-caused) activities can cause problems by concentrating the metals and their residues or slag into landfills that can leak and increase metal concentrations in water where they can become a problem for living organisms. So, while zinc and other metals may move from one part of the environment to another and even enter living organisms, these metals are not destroyed.

2.5.2   RADIOACTIVE ELEMENTS

Radioactive elements are those that can be naturally occurring such as C-14 or anthropogenic from nuclear reactors or similar sites. They have unstable nuclei and emit radiation such as x-rays. In the case of carbon, for example, 'normal,' nonradioactive carbon is C-12 with six protons and six neutrons. Carbon 14 has six protons and eight neutrons and emits these neutrons at a steady rate. Carbon 14 is formed by cosmic rays bombarding nitrogen in the atmosphere, resulting in instability. The half-life of C-14 is 5730 years, so, using a standard for C-14 assimilation of living organisms, scientists can carbon-date a fossil by determining the amount of C-14 left. For example, a fish fossil with 1/8 of the standard C-14 lived approximately 22,920 years ago. So, like metals, radioactive elements are not destroyed but their radioactivity spontaneously decays.

2.5.3   ORGANIC MOLECULES

There is another difference between metals and organic molecules. Organic molecules can be attacked by many agents and eventually degrade. Certainly, they can move about in ways similar to metals, but their overall concentrations will decrease. As we noted above, the rate of degradation varies highly from one organic contaminant to another and even among environmental compartments for the same molecule.

The primary factors causing the degradation of organic contaminants include biological organisms, ultraviolet radiation, spontaneous degradation, and weathering. Microbes such as bacteria and fungi break the chemical bonds between atoms and utilize the energy that exists in those bonds. For most organic molecules and microbes, the presence of oxygen can hasten the decomposition rates, so molecules that are tied up in deep aquatic sediments with no oxygen tend to persist for very long periods of time.

A problem inherent with many anthropogenic chemicals, however, is that they are **halogenated**. This means that the organic molecules are combined with elements such as chlorine, bromine, or fluorine. These elements increase the life span of the molecules in the environment in that the carbon–halogen bonds are very difficult to break. Molecules such as polychlorinated biphenyls (PCBs), dioxins, organochlorinated pesticides, and polybrominated diphenyl ethers (PBDEs) last for decades. Don't worry about these names for now; we will discuss them in greater detail later in this book. In contrast, other organic molecules such as many of our currently used pesticides and PAHs, which are not halogenated, have half-lives measured in days or a few months in soils and aquatic sediments.

Ultraviolet radiation or that portion of the electromagnetic spectrum that has shorter wavelengths than visible light (Figure 2.8) is a potent force for breaking down organic molecules. Sunlight is the source of this radiation, so molecules that are exposed to sunlight most often have shorter half-lives than the same ones that are shaded. Molecules in air or on the surface of water are most susceptible to ultraviolet radiation. Just as an example, Southworth (1979) estimated half-lives for the

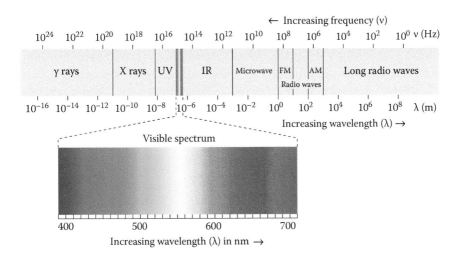

FIGURE 2.8 The electromagnetic spectrum in which visible light shares a place with radio waves, infrared (IR), ultraviolet (UV), and other types of radiation. (Courtesy of Wikimedia Commons, File:EM spectrum.svg, https://commons.wikimedia.org/wiki/File:EM_spectrum.svg, 2015.)

volatilization of anthracene (a polycyclic aromatic hydrocarbon) of 18 hours in a stream with moderate current and wind where the molecule was exposed to both oxygen and sunlight versus about 300 hours in a body of water at a depth of 1 m and no current but with less light and oxygen.

Many organic molecules can also spontaneously degrade over time due to their inherent instability. Weathering processes such as soil or sediment movements can aid in this degradation through physical means and by exposing the molecules to oxygen and light.

In addition to fate, we should include something about solubility indices when it comes to organic contaminants. Over the years, several methods of comparing solubility have been devised, but there are many different solvents in which organic chemicals can be exposed, some of which are very conducive to dissolving organic molecules, some not so much. In the environment, water is the principal solvent of concern, especially for aquatic ecotoxicology. However, many organic contaminants are hydrophobic, meaning that they do not mix well with water. For example, fresh oil will rise above and sit on top of water because it is hydrophobic ('water fearing'). As the oil weathers and decomposes, more of it will mix into the water column and its solubility increases. The reverse side of this coin is how contaminants interact with oil and oil-like liquids. Most of the major solvents for organic compounds are also organic, such as octanol, alcohol, and many others used in the laboratory. Molecules that dissolve easily in these solvents are said to be lipophilic ('lipid loving'). Not all contaminants are lipophilic; however, most metals, for instance, are more 'hydrophilic,' especially when the pH of the solution is in the acid range.

One direct method of determining water solubility is to determine how much of a known quantity of a contaminant will actually dissolve in water. One problem with this, however, is that water solubility is dependent on temperature—the warmer the water is, the greater the solubility. To control this over a broad range of conditions, chemists have developed the concept of octanol/water partitioning or K_{ow}. Following this concept, chemists, using sophisticated methods, have determined octanol/water partition coefficients (K_{ow}), which are defined as the ratio of a chemical's concentration in the octanol phase to its concentration in the aqueous phase of a two-phase octanol/water system. In other words, K_{ow} = Concentration in octanol phase/Concentration in aqueous phase. This differs from determining the solubility in octanol and water alone, for the biphasic nature of the solution involves some mixing of the solutes. Since concentrations are used in both the numerator and denominator of the equation, K_{ow} is unitless. Measured values of K_{ow} for organic chemicals have been found as low as 10^{-3} and as high as 10^7, thus encompassing a range of 10 orders of magnitude. In terms of log K_{ow}, this range is from −3 to 7. Thus, log values of K_{ow} are most often reported. Chemicals with low log K_{ow} values (e.g., less than 1) may be considered relatively hydrophilic; they tend to have high water solubilities and small bioconcentration factors for aquatic life. Conversely, chemicals with high log K_{ow} values (e.g., greater than 4) are very hydrophobic. Values between 1 and 4 are representative of chemicals with some degree of water solubility. Higher solubilities promote the uptake and loss of chemicals in living organisms, whereas moderate solubilities may lead to organisms having higher concentrations than environmental media and to increased concentrations in higher levels of food chains or webs.

Another partition coefficient of value to ecotoxicologists is the soil organic carbon/water partitioning coefficient or K_{oc}. This is the ratio of the mass of a chemical that is adsorbed in the soil per unit mass of organic carbon in the soil divided by chemical concentration in solution. For ecotoxicologists, the solution of interest is once again water. Note that the soil partition is standardized by the amount of organic carbon in that soil, for many contaminants adhere to organic carbons. K_{oc} values are useful in predicting the mobility of organic contaminants in soil; high K_{oc} values are associated with less mobile chemicals that tend to 'stick' to the soil, and this often reduces the bioavailability of the contaminants to living organisms. As K_{oc} values decrease, mobility increases, and a greater concentration of the compound enters the water column where it can become available to organisms. We will refer to these coefficients frequently through much of this book.

2.6   CHEMICALS OF MAJOR INTEREST

Although it is true that any chemical in the environment, if present in sufficiently high concentrations, may be toxic, ecotoxicologists have generally focused on certain chemical families. Chemicals in these families have been or are in locally high concentrations, dispersed in many parts of the world, usually long-lasting, and toxic even at parts-per-million concentrations or lower. Most also tend to be anthropogenic, although others occur naturally. We will cover these families in more detail later, but for now we will briefly describe them here.

2.6.1   POLYCHLORINATED BIPHENYLS AND SIMILAR CHEMICALS

PCBs used to be manufactured for a variety of purposes such as cooling fluids in electrical transformers. PCBs were discontinued in the 1970s. Similar chemicals with bromine called PBDEs and polybrominated biphenyls are still used as fire retardants in clothing and other materials. Dioxins and furans, which occur naturally through volcanic eruptions and combustion of organic matter at high temperatures, also have chemical similarities to these molecules.

2.6.2   POLYCYCLIC AROMATIC HYDROCARBONS

PAHs consist of two or more phenyl or benzene rings and only contain carbon and hydrogen atoms. They are frequently found in petroleum such as oil spills and can occur naturally.

2.6.3   ORGANOCHLORINE PESTICIDES

This group of pesticides included insecticides, herbicides, fungicides, and others. They all contained chlorine, and most were stable in the environment. Some could also biomagnify up food chains. The production and use of most of these have been banned in the United States and elsewhere, but a few are still used for limited purposes. They are very persistent in nature and are still found around the world.

2.6.4   CURRENTLY USED PESTICIDES

Actually, this is a huge group of chemicals from many different families. They are lumped into one category because to do otherwise could easily lead to an entire book just on these chemicals. They include insecticides, herbicides, fungicides, rodenticides, and many other chemicals for particular target organisms. In contrast to organochlorine pesticides, these chemicals do not contain halogens and are relatively short-lived. However, given the right conditions, some may persist for a year or more. The name might have been a giveaway, but these pesticides were initially produced to replace organochlorine pesticides in the 1970s, but since then, the group has become much more diverse and some of the original compounds have in turn been replaced by more efficacious ones.

2.6.5   METALS AND METALLOIDS

These naturally occurring elements can create contaminant issues due to artificially produced high concentrations from industrial applications, from combination with organic molecules, or in other ways. As mentioned above, only a few called heavy metals attract the greatest amount of attention, but any of them could potentially be problematic in the environment.

2.6.6   OTHER CHEMICALS

This is a hodgepodge of chemicals that do not fit easily into other groups but can still lead to environmental problems. They include plastics, radionuclides, munition compounds, and others.

2.7   UNITS OF MEASUREMENT

In ecotoxicology, scientists deal with very small concentrations of substances, often expressed as 1 millionth, billionth, or trillionth. Much of the time, water is the principal liquid medium in which contaminants are found, and concentrations are expressed as so many parts per liter (L) or milliliter (mL). Other times, contaminants are found in or on solids such as soil particles or living tissues, and concentrations may be expressed as parts per gram (g) or kilogram (kg). Similarly, the contaminants themselves may be in liquid or solid form. Prefixes for portions of liters or grams are milli- (1/1000, abbreviated m), micro- (1/1,000,000 or $1/1 \times 10^6$, abbreviated μ), nano- ($1/1 \times 10^9$, abbreviated n), or even pico- ($1/1 \times 10^{12}$, abbreviated p). Conversions from solid to liquid recognize that 1 g of water is equal to 1 milliliter (or mL). Thus, ecotoxicologists may speak of a concentration of 100 μL/L or

15 ng/kg, meaning 100 parts per million in a liter of water or 15 parts per billion per kilogram of soil.

Another factor to consider when looking at the concentration of contaminants in organisms is how the data are collected with regard to the organism itself. Often, scientists will report their findings as so many units of **live weight** or **fresh weight**. These terms are essentially synonymous and refer to the organism (e.g., a fish) that was freshly captured and sampled. **Wet weight** concentrations are similar, but some water loss may have occurred. At the opposite end of the scale, **dry weight** concentrations are taken after all moisture has been removed from the sample as through freeze-drying. Because animals may be 70%–85% + water and plant species can vary considerably in the amount of water they contain, dry weight concentrations should always be higher than fresh or wet weight concentrations, but the difference is going to vary from one organism to another. It is also very obvious that concentrations are going to differ according to tissue. In plants, roots often have higher concentrations of metals and metalloids because they absorb those constituents from the soil or sediment. Leaves, however, may have higher concentrations of pesticides and other organic chemicals than roots. Among animals, lipids, livers, and kidneys are often high in organic chemicals, but bone is a depository of lead and some other metals.

2.8 CHAPTER SUMMARY

1. Atoms are composed of electrons, neutrons, and protons with the number of protons identifying the specific element.
2. Electrons are arranged in shells or orbits around the nucleus; those farther away from the nucleus have greater energy than those closer to the nucleus.
3. Elements can be arranged into a Periodic Chart based on their number of protons.
4. Through donating, receiving, or sharing electrons, atoms can form reactions with each other to form molecules.
5. When one atom has a propensity to fully donate or fully accept electrons, they tend to form ionic bonds. Elements in the alkali metals and halogen classes most readily form ionic bonds. When ionically bonded molecules are placed in water, they tend to dissolve easily and disassociate into their respective ions.
6. Toward the center of the Periodic Chart, atoms have a strong tendency to share their electrons rather than fully donate or accept them. Sharing forms covalent bonds.
7. The two groups of chemicals of greatest interest to most ecotoxicologists are metals and organic molecules.
8. Metals are elements that can react with many other elements. For ecotoxicologists heavy metals (Cd, Co, Cr, Pb, Zn) are among the most important.
9. Organic molecules have carbon as their foundation. These molecules can be straight chains or circular in formation with one to three bonds between carbons.

10. The simplest organic molecules are composed only of carbon and hydrogen and are called hydrocarbons. Hydrocarbons can react with other molecules to form many of the environmental contaminants around our planet.

11. Metals and radioactive elements cannot be destroyed. However, they can move from one environmental compartment to another. In addition, radioactive elements and molecules spontaneously lose their radioactivity over time.

12. Organic molecules range from being short-lived to very persistent. Halogenation can greatly increase the persistence of organic molecules. The degradation of organic molecules can be hastened by biological organisms, ultraviolet radiation, oxygenation, or other factors.

2.9   SELF-TEST

1. Which subatomic particle defines an element?
   a. Neutron
   b. Proton
   c. Meson
   d. Electron

2. To have neutral charge, what two subatomic particles must be equal in number?
   a. Protons and neutrons
   b. Electrons and neutrons
   c. Protons and electrons
   d. Any of the above

3. What is the distinguishing feature of the elements in group 18 of the Periodic Chart?

4. What element has an atomic number of 6? How many total neutrons and protons does this element typically have?

5. True or False. Molecules formed by covalent bonds tend to disassociate into ions when dissolved in water.

6. Carbon is the foundation for organic molecules. What type of bond forms between carbon atoms?

7. True or False. Elements such as metals can be lost by changing into another element.

8. Which of the following factors often speeds up the process of degradation in organic molecules?
   a. Biological organisms
   b. Ultraviolet radiation
   c. Exposure to oxygen
   d. All of the above

9. A scientist wants to create a dose of a pesticide that has a concentration of 10 parts per million. How much pesticide should the scientist place in a vial containing 10 mL of water?
   a. 10 ng
   b. 100 µg

    c. 100 ng

    d. Because the pesticide is a solid and is being placed in a liquid, there is no way to determine how much pesticide should be added.

10. Which of the following structural formulas is a hydrocarbon?

    A.

    B.

    C.

    D.

    a. Only C is a hydrocarbon

    b. Only A is a hydrocarbon

    c. Only A and B are hydrocarbons

    d. All are hydrocarbons

REFERENCES

Agency for Toxic Substances and Disease Registry (ATSDR). 2003. *Toxicological Profile for Zinc*. Atlanta, GA: U.S. Department of Health and Human Services, Public Health Service.

Coffield, C. 2015. Period table of elements with free printable. Multiplication.com, http://www.multiplication.com/our-blog/caycee-coffield/periodic-table-elements-class-project-free-printable.

Los Alamos National Laboratory. 2016. Periodic chart of elements: LANL. http://periodic.lanl.gov/114.shtml. Accessed August 16, 2016.

Moore, J.T. 2011. *Chemistry for Dummies*, 2nd Edition. Hoboken, NJ: For Dummies.

Silberberg, M., Amateis, P. 2014. *Chemistry: The Molecular Nature of Matter and Change*, 7th Edition. Boston, MA: McGraw Hill.

Southworth, G.R. 1979. Role of volatilization in removing polycyclic aromatic-hydrocarbons from aquatic environments. *Bull Environ Contam Toxicol* 21: 507–514.

3 Some Ways Contaminants Affect Plants and Animals

3.1 INTRODUCTION

There are many, many ways contaminants can harm plants and animals. Sometimes they can inflict physical damage such as when ingested plastics block air passages or digestive tracts. Many contaminants affect enzyme production or functions and can interfere with photosynthesis or metabolism. Others cause cancer or result in malformations during development. Still others result in endocrine disruption (ED)—that is, they alter the synthesis or action of hormones. Some contaminants have been specifically designed to disrupt the nervous systems of animals that come in contact with them. In this chapter, we will examine some of these effects on the health of individual organisms and provide specific examples documented in the scientific literature.

3.2 GENERALIZED EFFECTS OF CONTAMINANTS

Among both plants and animals, the most general way that contaminants cause harm is by inhibiting their ability to thrive. In human medicine, the term 'failure to thrive' usually refers to infants or young children who do not gain weight or develop as quickly as the majority of the population. Often diagnosis is difficult to pinpoint. It is similar with plants and animals. Neither can tell an investigator how they are feeling, so we speak of **signs** rather than **symptoms** when the individual can relate how it feels. Animals, for example, cannot tell us if they have stomach aches or headaches and we have to make a reasoned guess as to what is happening internally, and that guess may not always be correct. Nevertheless, the overt effects of contaminant exposure are real.

For example, the harmful effects of heavy metals such as lead or copper on the growth of plants have been known for centuries (Prasad and Hagemeyer, 1999). In fish, preliminary signs of mercury poisoning include flaring of gills, increased respiratory movements, loss of balance, lots of mucous secretion, dark coloration, and sluggishness (Armstrong, 1979). Also, endosulfan, a pesticide that has been discontinued in the United States and elsewhere, reduces growth in Pacific tree frogs (*Pseudacris regilla*, Dimitrie and Sparling, 2014). The list of such observations is extensive, but the bottom line is that they do not require in-depth investigations into physiology—they can be recorded merely with the use of a balance or a ruler.

3.3 PHYSICAL OR MECHANICAL BLOCKAGE

A reasonably simple way in which contaminants affect organisms is through physical blockage of physiological processes. For instance, several sea animals such as turtles and seabirds may ingest plastics that float on the surface of the ocean. Plastic materials essentially do not degrade. Instead, these items break down into smaller and smaller pieces. These pieces can be consumed by fish and birds, which can block the gut, resulting in death. Dr Holly Gray et al. (2012) determined that plastic ingestion

FIGURE 3.1   Plastics and other materials collected from the digestive system of a Laysan albatross (*Phoebastria immutabilis*). (Courtesy of Siebel, E., Hier fressen die Vogel Kuntstoff, 2015, http://www.x-pansion.de/krass/detail/news/hier-fressen-die-voegel-kunststoff/, Accessed November 21, 2016.)

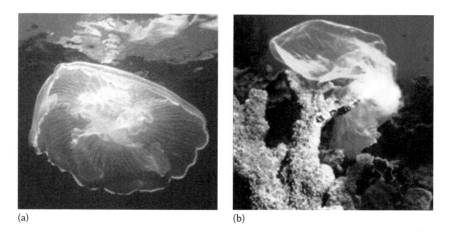

(a)   (b)

FIGURE 3.2   (a) Jellyfish. (b) Plastic bag. (Courtesy of McKitrick, K., Plastic bags and the environment, http://saynotoplastic.wikia.com/wiki/Plastic_bags_and_the_Environment, Accessed November 21, 2016.)

can be a serious factor for albatrosses in the North Pacific (Figure 3.1). Adult birds also feed ingested plastics to chicks which can cause problems. Other seabirds are similarly affected. Sea turtles that consume jellyfish, such as the loggerhead turtle (*Caretta caretta*), can mistake plastic bags for food items, which can also block the digestive system (Figure 3.2).

Some heavy metals such as cadmium, lead, and mercury accumulate in the liver and kidneys. Here they can block capillaries and ducts and interfere with normal organ functioning.

3.4 MALFORMATIONS

Strange body shapes can occur in animals due to contaminant exposure. If these unusual shapes occur during embryonic development, they are called **teratogenic**; but if they occur after the organism hatches or is born, they are collectively called malformations. Plants can also show weird growth forms such as twisting or stunting of some body parts because of contaminant toxicity.

Some brown bullheads (*Ameiurus nebulosus*) in the Anaconda River in Washington, DC, and the Detroit River in Michigan displayed large swellings on their face and tongues (Figure 3.3, Pinkney et al., 2001). Investigators have found that the tumors were caused by polycyclic aromatic hydrocarbons (PAHs) causing DNA adducts (see below) that led to cancerous tumors in adult fish.

In addition to stunted growth, Pacific tree frog tadpoles (*Pseudacris regilla*) exposed to endosulfan in the laboratory experienced a characteristic crook or bend in the spine just behind the head (Figure 3.4). Technically, this type of spinal malformation is called **scoliosis**. Scoliosis in these test animals caused them to swim in spirals and decreased food consumption, causing them to have smaller bodies at any stage

FIGURE 3.3 Brown bullhead (*Ameiurus nebulosus*) with tumor from the Anacostia River, Washington, DC. (Courtesy of Pinkney, F., U.S. Fish and Wildlife Service, https://www.fws.gov/chesapeakebay/newsletter/Spring12/Bullheads/Bullheads.html, Accessed November 21, 2016.)

FIGURE 3.4 Two larval Pacific tree frogs (*Pseudacris regilla*) of the same age. The one on the right is substantially smaller than the one on the left and has a decided crook or scoliosis in its body, which was caused by the pesticide endosulfan.

of development compared to controls. Although some lived as long as it took control animals to metamorphose into frogs, none of the tadpoles showed development (Dimitrie and Sparling, 2014).

In the 1980s, selenomethionine caused true teratogenesis in several species of birds nesting in the Kesterson National Wildlife Refuge. Drainage water from agricultural areas in the Central Valley of California brought an organic form of selenium, selenomethionine, into the refuge. Selenomethionine is a naturally occurring amino acid, but the concentrations found in water caused embryos to develop an assortment of teratogenic effects, including enlarged heads, very small or very large eyes, twisting or absence of beaks, and leg and wing malformations (Figure 3.5).

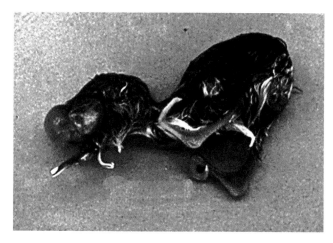

FIGURE 3.5 Deformed grebe embryo collected from selenomethionine-contaminated Kesterson National Wildlife Refuge a few decades ago. (Courtesy of U.S. Geological Survey.)

Refuge managers blocked incoming water from the refuge, filled in wetlands, and closed much of the refuge down. A noncontaminated portion of the old refuge was later incorporated into the San Luis National Wildlife Refuge (Ohlendorf, 2011).

3.5 INTERFERENCE WITH ENZYME ACTIVITY

3.5.1 CHELATION

The physiology of any organism is modulated by hundreds to thousands of enzymes that facilitate chemical reactions within the body. Enzymes are present in cells and in the fluid portion of blood called plasma. Enzymes operate in a lock-and-key manner in which the shape of the enzyme is specific for a particular reactive site on another molecule. These enzymes facilitate chemical reactions in organisms by lowering the amount of energy necessary to start reactions. Interfering with enzyme-facilitated reactions is a common way in which contaminants produce harmful effects. Many contaminants, most notably heavy metals, interfere with these enzymes in several ways. One method is **chelation**, in which the metal binds with the enzyme, alters its shape, and prevents its ability to carry out the reaction.

3.5.2 RECEPTOR BINDING

Another way that contaminants affect enzyme activity is to bind with receptors on the surface of cells. In an overly simplified explanation, certain molecules or receptors are situated on the surface of cells, waiting for an appropriate chemical signal to come by. The receptors are usually highly specific for certain molecules called ligands, again in a lock-and-key fashion based on the molecular shapes. These ligands are often enzymes, hormones, or other proteins. In one type of reaction, when the appropriate ligand binds with the receptor (Figure 3.6), the receptor facilitates the movement of the ligand across the cell membrane and various cellular activities can occur. For instance, the porosity of the cell membrane may change to allow or block molecules

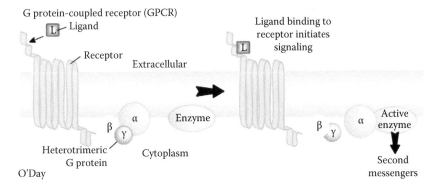

FIGURE 3.6 Graphic representation of the binding of ligand, which might be a contaminant, with receptors on the surface of a cell membrane resulting in several intracellular reactions. (Courtesy of O'Day, D.H., *Introduction to the Human Cell: The Unit of Life and Disease*, University of Toronto Mississauga, Mississauga, Ontario, Canada, 2010.)

entering the cell. Also, the receptor–ligand bond may invoke energy in the form of ATP to be released, thereby facilitating chemical reactions. Other enzymes and molecules may be involved in the total picture. Some of these may be released into the cytoplasm and diffuse to the nucleus where they can alter RNA replication. This in turn can affect the production of other proteins. We simply cannot go into the full details necessary to address all of the possibilities. Suffice to say, contaminants may also bind with some receptors, thereby affecting their functions by stimulating them, inhibiting them, or blocking them from reacting with the appropriate ligand.

3.5.3 NEUROTRANSMITTER INTERFERENCE

Other enzyme-mediated reactions that can be affected by contaminants are associated with the nervous system. One such process involves the enzyme **acetylcholinesterase** (AChE) at the synapse between two neurons or between a neuron and a muscle. When nerve cells connect with muscle fibers to innervate them, there is a space or synapse between the nerve cell and the muscle fiber (Figure 3.7). While the nerve cell transmits an electrical impulse through its entire length, the electrical impulse cannot be transmitted across this synapse. Instead, the impulse triggers the release of **acetylcholine** (ACh), which diffuses across the synapse or the synaptic cleft. If sufficient molecules reach the postsynaptic membrane, the motor fiber is stimulated to contract. As long as there is ACh in the synaptic cleft, additional motor fiber contractions can occur. To prevent this from happening, the nerve fiber secretes the enzyme AChE into the cleft, which breaks down ACh and prevents uncontrolled

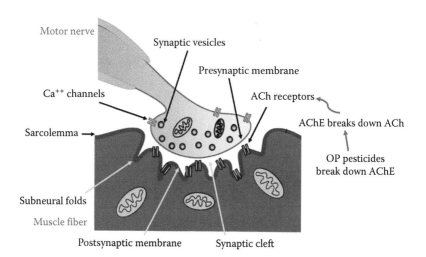

FIGURE 3.7 Graphic representation of a neuron-muscular junction with synapse. Organophosphorus and carbamate pesticides function by deactivating the enzyme acetylcholinesterase (AChE), which breaks down the neurotransmitter acetylcholine. With AChE gone, the neurotransmitter stays in the synaptic cleft and causes continual firing of the muscle, which ultimately can cause death through asphyxiation. (Courtesy of Wikimedia Commons, Neuromuscular junction, https://en.wikipedia.org/wiki/Myasthenia_gravis, 2017.)

muscular spasms. ACh is also active in the brain. Many modern pesticides have been designed to make use of this system. They react with AChE and block its ability to break down ACh. Consequently, the muscle fibers contract in erratic and uncontrolled ways; death is usually due to asphyxiation when the animal can no longer regulate breathing. These pesticides can also result in abnormal behavior. The most potent inhibitors of AChE are organophosphorus insecticides, including carbamates and organophosphates. Pyrethroids also inhibit AChE (see Chapter 5 for a more thorough explanation of these chemical families), and some other pesticides affect ACh as a side effect.

Other neurotransmitters may be affected by pesticides. For instance, gamma-aminobutyric acid (GABA) is an amino acid that acts as a neurotransmitter in the central nervous system. Rather than working at the synapse, however, GABA operates along the nerve axon by modulating a 'gate' that allows the flow of chlorine into and out of the nerve cell. Some organochlorine pesticides such as endosulfan, dieldrin, and lindane interfere with GABA, causing neurons and muscle contraction to be hyperexcited. Death is also usually due to asphyxiation.

3.5.4 Effects of Contaminants on Plants

We should not go on to the next section without at least a word about plants. Many studies have been conducted on the effects of heavy metals and other contaminants on plants. Do we need to mention that herbicides are widely used in developed nations? In part, these studies have mainly examined the effects of pollutants on photosynthesis and other physiological processes, and in part they have been conducted to identify plants that might be effective in the remediation of contaminated sites. If plants can take up contaminants through their roots, they can then be harvested and disposed of safely, thus reducing the amount of contaminant in the soil. Many of these studies (e.g., Beals and Byl, 2014; Deng et al., 2014) have shown that a variety of contaminants can affect enzyme activity, often resulting in decreased photosynthesis. Several herbicides that are used commercially, including the world's most widely used herbicide glyphosate, operate by interfering with enzyme activity in plants.

3.6 IMPINGEMENTS ON THE IMMUNE SYSTEM

An animal's immune system is very complex. It starts with nonspecific factors that protect against any foreign invader. The first line of defense is the skin, followed by stomach secretions, and mucous of epithelial tissues. Also included in this nonspecific category are white blood cells or leukocytes in the circulatory system that ingest bacteria and small parasitic organisms. Following this protection, the body has an adaptive, specific system of protection, which operates by learning to recognize and attacking foreign antigens.

All foreign organisms, including viruses, have proteins on their surfaces; these proteins can serve as antigens. When foreign agents such as toxins, bacteria, or viruses enter another organism, lymphocytes in the host become sensitized. Two types of lymphocytes, B cells and T cells, exist, with B cells coming from the bone marrow (in mammals) or liver and T cells from the thymus. It takes some time for these lymphocytes to produce their specificity, so the first encounter with foreign antibodies may result in a slow response. However, once the system has been activated, sensitized lymphocytes are continually produced so that if subsequent exposures occur, the host organism can respond more quickly.

When activated, B cells form antibodies, which are small proteins that attack the antigens. Antibodies can neutralize some toxins, deactivate viruses, activate cell-killing proteins called **complement** that work with the antibodies to destroy cells, or attract leukocytes that engulf the foreign agents. Activated T cells will attack a foreign substance with their entire bodies or make substances called lymphokines that signal to leukocytes to go to the site of infection. The National Institutes of Health (NIH, 2013) has published an online discussion on human immunity that goes into much more detail than can be covered here.

Contaminants can potentially alter animal immune functions in many ways. However, the interaction between ecotoxicology and epidemiology is not as well established as it might be. In general, scientists studying wildlife diseases and those studying the effects of contaminants need to collaborate more formally; each tends to focus on their specific areas without much integration. The marriage of these two disciplines has been called **immunotoxicology**, and this field has focused more on human health than on wildlife or plants. Even within animals, most studies have been on laboratory rats and mice. Much of the work on plants has focused on the influence of nutrition on human immunotoxicology. Nevertheless, studies have shown that contaminants can cause cellular damage in organs responsible for immune function, such as the thymus. Experimental exposures to pesticides and other chemicals have reduced lymphocyte numbers and induced structural changes in T cells (Galloway and Handy, 2003). Because contaminants interfere with enzymes in many ways, they could also affect enzymes that are involved with immune functions. This could interfere with the signaling that goes on between T cells and leukocytes. Several studies have shown that low doses of contaminants can actually increase immune functions, whereas higher doses decrease overall immunity (e.g., Kreitinger et al., 2016).

Heavy metals, particularly cadmium, lead, and mercury, also negatively impact immune functions in laboratory animals. Exposure to environmental concentrations similar to those found in fish from the Great Lakes showed that these metals lead to increased susceptibility to infections, autoimmune diseases, and allergic manifestations (Bernier et al., 1995). As we will see in a little bit, several halogenated organic contaminants can cause cancer. An unanswered question so far is whether contaminant exposure under natural conditions can lead to decreased immunity that could influence population dynamics.

3.7   CONTAMINANT-INDUCED ENDOCRINE DISRUPTION

The endocrine system consists of several glands that do not have direct connection to the circulatory system and secrete hormones. These hormones control many processes in an organism, including metabolism, reproduction, responses to stress, and functioning of the nervous system. Any alteration in the normal operation of the endocrine system can be called **endocrine disruption**. ED can occur through many routes, both natural and man-made, but here we will focus on how contaminants cause problems.

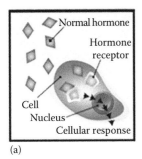 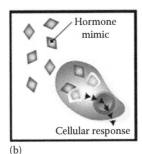 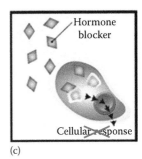

(a)　　　　　　　　(b)　　　　　　　　(c)

FIGURE 3.8 Representation of how a contaminant can serve as an endocrine disruptor. A normal response occurs between a hormone and the cell (a). As a hormone mimic (b), the contaminant produces a similar response as the natural functioning hormone. As a blocker, (c) it prevents the hormone receptor from interacting with the hormone and ceases the cellular response. (Courtesy of National Institutes of Health (NIH), Endocrine disruptors, http://www.niehs.nih.gov/health/topics/agents/endocrine/.)

Contaminants that function as endocrine disruptors can (1) mimic naturally occurring hormones in the body, like estrogens (the female sex hormone), androgens (the male sex hormone), and thyroid hormones, potentially producing overstimulation; (2) bind to cellular receptors and block the naturally occurring hormones from binding and the normal signal then fails to occur and the body fails to respond properly; and (3) interfere or block the way natural hormones or their receptors are made or controlled, for example, by altering their metabolism in the liver. Some organic contaminants that exert ED effects include phthalates, bisphenol A (both associated with plastics), dioxins, furans, pharmaceuticals, dichlorodiphenyltrichloroethane (DDT), polychlorinated biphenyls (PCBs), and some pesticides.

A simple representation of how ED works is shown in Figure 3.8. In a healthy cell, hormones trigger specific receptors on the cell membrane. The stimulus then either allows the hormone to pass into the cell or the cell releases intrinsic chemicals. Either way, when these reach the nucleus, they cause certain portions of the DNA to unravel, allowing transcription of RNA and subsequently synthesis of enzymes or other proteins. ED mimics cause the cell to behave in the same way as the respective hormone, leading to protein synthesis, even though the proteins are not needed or wanted. ED blockers prevent the normal processes from occurring and the required proteins are not produced.

3.8 GENOTOXICITY

Any alternation in the genetic makeup of an organism caused by contaminants or other factors such as radiation constitutes **genotoxicity**. These alterations can include simple mutations in a sequence of nucleotides making up a strand of DNA such as substitutions of one nucleotide by another, insertions of one or more new nucleotides, reversals in a segment of a strand, or deletions in which one or more nucleotide is removed from the sequence. The alteration can also include chromosomal damage such as breakage. Genotoxicity can occur in a wide range of species, both plant and animal, and with many different contaminants including heavy metals, PAHs, organochlorine pesticides, and PCBs. Fortunately, I suppose, most of the research has shown that genotoxicity occurs mostly in somatic cells—that is, body cells such as liver, kidney, and skin. These mutations might impact physiological processes in an organism and even lead to cancer, but they cannot be transmitted to the next generation. Substantially fewer studies have shown mutations in germ cells—sperm or eggs. These might be passed on to subsequent generations. Certain contaminants such as benzo[a]pyrene, a PAH (e.g., Godschalk et al., 2015), and radionuclides (e.g., Pomerantseva et al., 1997) are well known for their ability to induce mutations in spermatocytes of laboratory rodents.

There are several ways in which contaminants can cause genotoxicity, but one well-studied way is through the formation of **adducts**. Contaminants that enter a cell by one of the methods described above can attach to DNA or RNA molecules. The union forms the adducts. Such modified DNA may replicate erratically, providing false information to the cell, which can disrupt its functions. PAHs and other petroleum derivatives are well known for their formation of adducts. As above, among the most toxic of these is benzo[a]pyrene, which is very carcinogenic (White and Claxton, 2004).

3.9   CYTOCHROME P450 AND OTHER METABOLIC SYSTEMS

Under normal conditions animals are not defenseless against toxins. Many by-products of metabolism are toxic at sufficient concentrations. Often these could produce **oxidative stress**. Oxidative stress is due to an imbalance between the production of free radicals and the ability of the body to counteract or detoxify their harmful effects through neutralization by antioxidants. A free radical is an oxygen-containing molecule that has one or more unpaired electrons, such as peroxide (O_2^{-2}) and superoxide (O_2^{-1}), making it highly reactive with other molecules. Free radicals can chemically interact with cell components such as DNA, proteins, or lipids and destabilize them. They in turn seek and steal an electron from another molecule and trigger a large chain of free radical reactions. Oxidative stress eventually can destroy cells or alter their function and has been linked to several diseases, including cancer, muscular dystrophy, autoimmune diseases, emphysema, Parkinson's disease, multiple sclerosis, and atherosclerosis.

There are systems intrinsic to cells that have evolved to protect the cells from potentially harmful natural chemicals. One major system is called Cytochrome P450. Cytochrome P450 is a complex system consisting of cytochromes, which are iron-containing hemoproteins (proteins attached to iron atoms). In plants, cytochromes are instrumental in photosynthesis, and in both plants and animals, other cytochromes are essential in conducting particular chemical reactions. The chief cytochrome system in breaking down contaminants is the very complex Cytochrome P450 containing thousands of enzymes. A thorough discussion of this complex system is too much for this introductory book, but interested students might check out an informative series of e-articles by McDowell (2006).

A problem arises when many organic contaminants meet the P450 system. These are man-made chemicals that are foreign to P450. Often, these chemicals are treated adversely, and more free radicals are produced rather than destroyed. As a result, the toxicity of the parent compounds can actually increase due to their metabolism by P450. For instance, PCBs are often found in their hydroxylated form in the environment, PCB-OH. When metabolized by P450, the hydroxyl becomes a free radical that can induce oxidative stress on top of the underlying toxicity of the PCB itself. Dioxins, benzo[a]pyrene, and some of the most toxic organic contaminants behave similarly.

3.10   SO HOW DO ECOTOXICOLOGISTS STUDY THESE EFFECTS?

We have covered only some of the many harmful effects that environmental contaminants can inflict on living organisms. Much of what we have included has been on multicellular animals, but contaminants do not discriminate. Microbes, plants, animals, and humans can be negatively affected by these chemicals. Many of the effects have been identified under laboratory conditions where all outside factors can be controlled. Less is known about how these effects might express themselves in natural environments. Massive die-offs of organisms due to unplanned exposure to contaminants are seldom seen. Instead, contaminants often affect organisms and populations more subtly, and populations' effects have been notoriously difficult to identify.

Let us leave this list of specific effects and discuss how scientists determine what effects occur in the laboratory. In the early days of ecotoxicology, back in the 1970s, major concern focused on how much of a contaminant was needed to kill selected organisms. At first, scientists used a variety of methods and organisms for their testing. Soon, however, the U.S. Environmental Protection Agency (EPA) began to formalize a methodology. One of the earliest standardized methods, called a **protocol**, was the determination of the acute median lethal dose (or concentration). An acute test is one that takes place over a short period of time, and protocols were developed for 24, 48, and 96 hour exposures. The principal objective was to establish the **median lethal dose** or **LC50** (C stands for concentration). This is the estimated concentration at which 50% of a test population is expected to die.

In brief, a scientist (or his technician or graduate student) sets up an experiment consisting of several predetermined concentrations of the contaminant of interest and establishes several replicates per concentration. Each replicate consists of 1 to several exposed organisms. Control organisms that are treated in exactly the same way as test subjects, except that they are not exposed to the contaminant, are a critical part of these studies. Through time, a battery of preferred species was selected to make these tests even more standardized. For aquatic tests, the principal subjects often included plants such as *Elodea* sp., invertebrates like the crustaceans *Hyalella* sp. or *Daphnia* sp., the sediment-dwelling annelid *Lumbriculus variegatus*, fish species such as rainbow trout (*Oncorhynchus mykiss*) or fathead minnow (*Pimephales promelas*), and the African clawed frog (*Xenopus laevis*) (Figure 3.9). Other species that inhabited polluted waters are used if they have ecological importance. Typically, one test contaminant is used at a time by dissolving it in water. The organisms are observed daily for the duration of the experiment and then removed

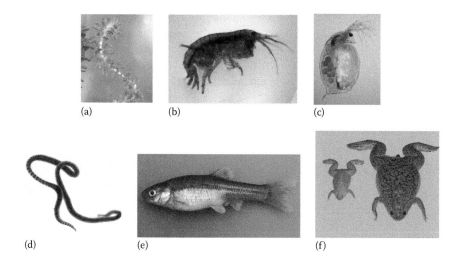

(a)   (b)   (c)

(d)   (e)   (f)

FIGURE 3.9   Examples of common species used in aquatic toxicity testing: (a) *Elodea* sp., (b) *Hyalella azteca*, (c) *Daphnia magna*, (d) *Lumbriculus variegatus*, (e) *Pimephales promelas*, and (f) *Xenopus laevis*.

at the end of the stipulated period and examined for any obvious abnormality, although the number of survivors at the end of the test is the primary concern. The resulting data can be graphed (Figure 3.10) and the LC50 can be determined. The relative toxicity of two or more compounds may then be compared, or the information can be used in risk assessment (see Chapter 11). More precisely, the

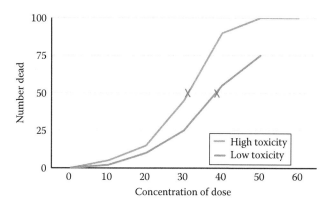

FIGURE 3.10   Dose/response curves can be drawn by plotting the cumulative number of organisms affected against the dose of the contaminant. The orange curve represents a contaminant with a lower toxicity than the blue curve. The higher median effective dose (LD50 or LC50) when death is used as the end point is 40 units for the orange and 30 for the blue, which means that it would take 40 units of orange to produce the same effect as 30 units of blue.

data are statistically analyzed by specific methods that yield both an estimate of the LC50 (or any other point on the graph) and confidence intervals around the LC50 to provide an estimate of error.

Another approach is to work with terrestrial species. Here, appropriate housing such as indoor or protected outdoor cages is designed. Traditional test animals include northern bobwhite (*Colinus virginianus*), mallard (*Anas platyrhynchos*), chicken, or various laboratory mammals such as rabbits, rats, and mice. Animals may be injected or gavaged with the known concentration of the contaminant. Gavage is a method that usually inserts a tube down an animal's esophagus and places a gelatin capsule through the tube into the animal's stomach or gizzard. Because the concentration is known, these methods also result in LC50 information. Another method often used with waterfowl or other birds is to mix the contaminant into their food. Birds can be messy eaters and spill some of their food, so the exact ingested concentration is unknown. However, this type of exposure does a good job in representing what birds might be picking up while foraging. The term for the results of this type of exposure is **LD50** with D standing for dose.

LC50 and LD50 tests are still often used today. However, field data have shown that (1) organisms are often exposed to contaminants for more than 96 hours, often for days, or even lifetimes; (2) exposure concentrations are often much lower than those used in median lethal tests; and (3) sublethal effects are often observed well before mortality is witnessed. To enhance the realism in these laboratory studies, scientists have looked at chronic exposures and subtler effects than dead animals lying at the bottom of their cage or aquarium. Chronic exposures occur over days to months. The U.S. EPA has developed a protocol for a 28-day exposure, but even this may not be long enough, depending on the nature of the organism or expected effects. Here is where scientists working on specific problems can be more creative. I have often exposed tadpoles of various species to low concentrations of pesticides or other contaminants from the time they hatch out to when most have metamorphosed into juvenile frogs. These tests mimic the duration of exposures under real-world scenarios. Alternatively, researchers may be interested in whether a contaminant produces cancer, ED, or some other malady and have to resort to exposures that are long enough for these effects to develop.

Laboratory exposures can be tightly controlled and are appropriate for addressing basic problems in ecotoxicology. However, organisms seldom live under controlled day length, constant temperatures, absence of predators, or other refinements seen in captivity. As a result, more recent studies in aquatic species are being conducted in large outdoors, stock tanks, or small swimming pools called **mesocosms** (Figure 3.11) that allow interactions among contaminants and other environmental factors. Similarly, efforts to make more realistic exposures in terrestrial animals are being considered but are not as far along as those in aquatic toxicology. An alternative is to construct cages in the actual polluted environment, a method called *in situ* experiments. In addition, because organisms most often live in a 'cocktail' or 'soup' of contaminants, studies are beginning to use more than one contaminant and looking for possible interactions.

FIGURE 3.11   Small swimming pools and stock tanks are being used to provide aquatic organisms with a greater realism than aquaria in laboratory setting. (Courtesy of WikiMedia Commons by Freddyfish4, https://commons.wikimedia.org/wiki/File:Cattle_tank_mesocosm_array.JPG)

3.11   CHAPTER SUMMARY

1. A major focus of ecotoxicology is how contaminants affect living organisms.
2. There are many possible effects that can result from exposure to contaminants; death is one among them.
3. Plants and animals cannot tell investigators how they are feeling or where they hurt. Therefore, ecotoxicologists have to rely on signs rather than symptoms and use indirect ways of assessing the health status of test animals.
4. Animals may be affected by contaminants in many ways:
    a. Generalized signs of contaminant exposure may be lethargy, lack of appetite, hiding, reduced growth, or diminished vigor.
    b. One of the simplest effects of contaminants is blockage of digestive systems, organs, or circulatory systems.
    c. Malformations may occur due to contaminant exposure. If they occur prior to birth or hatching, the circumstance is called teratogenic.
    d. Many contaminants, including heavy metals, can interfere with normal enzyme functioning. They can do this by chelating the enzyme, interfering with its synthesis, or blocking it from exerting its effect. One example of this is the inhibition of AChE in the nervous system by organophosphorus, carbamate, and pyrethroid pesticides.
    e. Potentially, contaminants can interact with the immune systems of plants and animals to produce immunotoxicity. Laboratory studies and some field investigations report that all aspects of the immune system are subject to harmful effects of contaminants. Whether these effects result in decreased survival under natural conditions is not well known.
    f. ED occurs when contaminants impede the normal functioning of the endocrine system. In some cases, contaminants can inhibit hormonal activity; in others, they can mimic the activity.
    g. Some contaminants have been found to cause mutations in living organisms. These mutations can be at the gene or chromosomal level. Again, it is not well known if these mutations can affect populations.

5. Plants and animals have developed several systems that protect them from internal toxins. Chief among these is the P450 system. Man-made contaminants, however, can actually become more toxic when they are metabolized by the P450 cytochromes.
6. A common objective in toxicity testing has been to determine the lethal median dose (LC50 or LD50) under acute exposures. Today, toxicity testing has expanded to identify sublethal effects through the use of chronic exposures. Outdoor laboratories and the use of more than one contaminant at a time have increased the realism in toxicity testing.

3.12   SELF-TEST

1. Why do ecotoxicologists rely on signs rather than symptoms in studying the effects of contaminants on organisms?
2. Provide at least one example of how contaminants can physically block passages or organs.
3. What is the difference between teratogenic effects and other kinds of malformations?
4. How can metals and other contaminants affect enzymes?
   a. Through binding with them and changing their shape in a process called chelation
   b. By blocking their reactive sites on the surface of cells
   c. By causing genetic effects that distort the synthesis of proteins
   d. All of the above
5. What enzyme is present in the synapse between nerve and muscle cells and is blocked by certain pesticides?
   a. Endosulfan
   b. Acetylcholine
   c. Acetylcholinesterase
   d. Hyaluronidase
6. True or False. It has been well documented in several species that contaminants can cause serious population effects because of their damage to immune systems.
7. Which of the following contaminants affects the hormones that determine sexual development?
   a. Bisphenol A
   b. Benzo[a]pyrene
   c. Carbamate
   d. None of the above
8. True or False. The Cytochrome P450 system provides a reliable and safe way of breaking down all man-made contaminants in organisms.
9. What is the difference between an LC50 and an LD50?
10. In comparing toxicity tests under laboratory and field conditions, which of the following statements is true?
    a. Laboratory studies offer greater control during exposures.

b. Laboratory studies can be modified to very accurately simulate field conditions.

c. By their very nature, field studies offer greater randomness than laboratory studies.

d. More than one above.

REFERENCES

Armstrong, F.A.J. 1979. Effects of mercury compounds on fish. Pp. 657–670. In: Nriagu, J.O. (ed.), *The Biogeochemistry of Mercury in the Environment*. New York: Elsevier/North Holland Biomedical Press.

Beals, C., Byl, T. 2014. Chemiluminescent examination of abiotic oxidative stress of watercress. *Environ Toxicol Chem* 33: 798–803.

Bernier, J., Borusseau, P., Krzystyniak, K., Tryphonas, H., Founier, M. 1995. Immunotoxicity of heavy metals in relation to Great Lakes. *Environ Health Perspect* 103(Suppl 9): 23–34.

Deng, G., Ming, L., Hong, L., Yin, L.Y., Li, W.L. 2014. Exposure to cadmium causes declines in growth and photosynthesis in the endangered aquatic fern (*Ceratopteris pteridoides*). *Aquat Bot* 112: 23–32.

Dimitrie, D.A., Sparling, D.W. 2014. Joint toxicity of chlorpyrifos and endosulfan to Pacific treefrogs (*Pseudacris regilla*) tadpoles. *Arch Environ Toxicol* 67: 444–452.

Galloway, T., Handy, R. 2003. Immunotoxicity of organophosphorus pesticides. *Ecotoxicology* 12: 345–363.

Godschalk, R.W.L., Verhofstad, N., Verheijen, M., Yauk, C.L., Linschooten, J.O. et al. 2015. Effects of benzo[a]pyrene on mouse germ cells: Heritable DNA mutation, testicular cell hypomethylation and their interaction with nucleotide excision repair. *Toxicol Res* 4: 718–724.

Gray, H., Lattin, G.L., Moore, C.J. 2012. Incidence, mass and variety of plastics ingested by Laysan (*Phoebastria immutabilis*) and Black-footed Albatrosses (*P. nigripes*) recovered as by-catch in the North Pacific Ocean. *Marine Poll Bull* 64: 21980–21992.

Kreitinger, J.M., Beamer, C.A., Shepherd, D.M. 2016. Environmental immunology: Lessons learned from exposure to a select panel of immunotoxicants. *J Immunol* 196: 3217–3225.

McDowell, J. 2006. Cytochrome P450. Interpro. http://www.ebi.ac.uk/interpro/potm/2006_10/Page1.htm. Accessed December 1, 2016.

McKitrick, K. Plastic bags and the environment. 2013. http://saynotoplastic.wikia.com/wiki/Plastic_bags_and_the_Environment. Accessed November 21, 2016.

National Institutes of Health (NIH). 2017. Endocrine disruptors. http://www.niehs.nih.gov/health/topics/agents/endocrine/. Accessed April 19, 2017.

National Institutes of Health (NIH). 2013. Overview of the immune system. https://www.niaid.nih.gov/topics/immunesystem/Pages/overview.aspx. Accessed November 21, 2016.

O'Day, D.H. 2010. *Introduction to the Human Cell: The Unit of Life and Disease*. University of Toronto Mississauga, Mississauga, Ontario, Canada.

Ohlendorf, H.M. 2011. Selenium, salty water, and deformed birds. Pp. 325–357. In: Elliott, J.E., Bishop, C.A., Morrissey, C.A. (eds.), *Wildlife Ecotoxicology: Forensic Approaches*. New York: Springer.

Pinkney, A.E., Harshbarger, J.C., May, E.B., Melancon, M.J. 2001. Tumor prevalence and biomarkers of exposure in brown bullheads (*Ameiurus nebulosus*) from the tidal Potomac River, USA, watershed. *Environ Toxicol Chem* 20: 1196–1205.

Pinkney, F. 2012. Solving the riddle of skin tumors in brown bullhead catfish. U.S. Fish and Wildlife Service. https://www.fws.gov/chesapeakebay/newsletter/Spring12/Bullheads/Bullheads.html. Accessed November 21, 2016.

Pomerantseva, M.D., Ramaiya, L.K., Chekhovich, A.V. 1997. Genetic disorders in house mouse germ cells after the Chernobyl catastrophe. *Mut Res Fund Mol Mech Mutagen* 381: 97–103.

Prasad, M.N.V., Hagemeyer, J. 1999. *Heavy Metal Stress in Plants: From Molecules to Ecosystems*. New York: Springer.

Siebel, E. 2015. Hier fressen die Vogel Kuntstoff. http://www.x-pansion.de/krass/detail/news/hier-fressen-die-voegel-kunststoff/. Accessed November 21, 2016.

White, P.A., Claxton, L.D. 2004. Mutagens in contaminated soil: A review. *Mut Res Fund Mol Mech Mutagen* 567: 227–345.

Section II

Major Groups of Contaminants: Where They Come from, What They Can Do

4 Metals

4.1 INTRODUCTION

In the next few chapters, we will study organic contaminants, those that are mainly man-made and have carbon as their principal constituent. Most of these compounds were specifically designed for their ability to either kill or repel organisms. In this chapter, we discuss a very different type of contaminant that is both inorganic and, in some cases, actually required for life. These are the elements collectively called metals that occupy most of the Periodic Chart of Elements (see Figure 2.2).

Given the number of elements that belong to the metals group, you can imagine that there is a wide variety in their chemical behaviors. Metals at the far left of the Periodic Chart are highly soluble in water to the point that they need to be protected from water. For instance, sodium is highly explosive when it comes in contact with water. As we progress to the right of the chart, we encounter metals that are soluble in slightly acidic water to those that are only soluble in very acidic solutions. Mercury does not dissolve in water, although a temporary suspension can be made with this heavy metal. Metalloids share some but not all of the properties of metals, and those of major interest to the ecotoxicologist include boron (B), selenium (Se), germanium (Ge), arsenic (As), actinium (Ac), tellurium (Te), and polonium (Po).

Perhaps this is a good place to clarify some terms. Environmental chemicals including contaminants are taken up into the body or **assimilated** when they become incorporated into tissues. When chemicals leave the body through urination, defecation, or respiration, or are broken down into simpler compounds that leave the organism, they are said to have been **depurated**. Some contaminants including heavy metals may have higher rates of assimilation than depuration and thus **bioconcentrate**, meaning that the concentration of a particular compound is higher than in the environment. If the concentrations of a compound increase up the food chain due to transfer from one level to another, they are said to **biomagnify**. Except for mercury or when they are combined with organic molecules and become **organometals**, heavy metals seldom, if ever, biomagnify. Some contaminants including organometals are both **lipophilic** and **hydrophobic**, however, in that they concentrate in fats within organisms and do not mix well with water. In contrast, other molecules are both **lipophobic** and **hydrophilic** if they avoid lipids and mix well or dissolve readily in water.

4.2 SOURCES OF METALS IN THE ENVIRONMENT

The lithosphere or terrestrial environment is the principal reservoir of metals (Figure 4.1). Metals are located at all levels of the earth, from surface soils to magma in its core. Those close to the surface can be blown around in the atmosphere or washed into bodies of water in runoff. Earthquakes, upheavals, and especially volcanoes bring metals from deep within the earth to the crust. Metals in the atmosphere

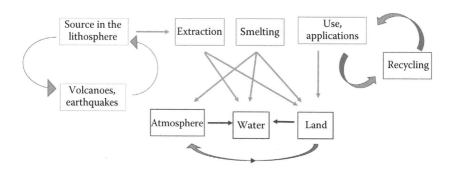

FIGURE 4.1   Major sources of metals in the earth and how the different sources interact.

may be carried to remote areas until they are deposited back onto land or water. Metals in water may be deposited into sediments for eons or carried to oceans.

Metals with economic value are mined and smelted to produce purer forms of the raw substance. These activities have been principal anthropogenic sources of metals in the environment. Other sources of environmental contamination occur whenever metals are used in industrial applications. Prior to catalytic converters, automobile exhausts were also significant sources of lead, but today catalytic converters require unleaded gasoline. Combustion of coal can also be a major source of metal pollution into the atmosphere. Such pollution continues today but many developed countries have strict regulations to reduce the output of these metals by requiring smokestack scrubbers and other technologies to clean the exhausts before they leave the stack. Many developing countries do not have such regulations, and metal pollution remains a global problem.

Just where metals are distributed and how much depends, of course, on the metal. Table 4.1 provides some concentration data for major environmental compartments

TABLE 4.1

Concentrations of a Few Heavy Metals in Various Environmental Compartments along with Annual Production Amounts

Metal	Earth's Crust (mg/kg)	Freshwater (µg/L)	Marine (µg/L)	Atmosphere (ng/m³)	Annual Production Amount (tonnes[a])
Cadmium	0.5	0.05	5.5	0.03	24,000
Chromium	125	10–100	0.3	1.0	748 million
Copper	2–250	1–50	0.06–6.7	1–3	720 million
Lead	16	2	0.02	3.5	89 million
Mercury	0.67	0.0004–0.0015	<0.0002	1–20	1,870
Nickel	75	7	0.0006	1–15	79 million
Zinc	40	0.02–0.04	0.003–0.007	0.001–0.04	200 million

Note: Values are from uncontaminated sites and can be orders of magnitude higher for contaminated areas. Data have been taken from several sources.

[a]   A tonne or metric ton is equivalent to 1000 kg or 1.1 English tons.

and a handful of metals. As you can see, the earth's crust has the highest concentrations. This is followed by water—either fresh or marine—and finally by air. The table also includes data on the annual extraction and refinement of these metals. As might be expected, copper and chromium lead the list in this category, but neither of these even compare to the big production metals such as aluminum or iron, which are extracted at the trillions of tonnes level.

4.3 FACTORS AFFECTING THE BEHAVIOR OF METALS

Metals can occur in their inorganic forms either as ions or as nonionic elements. They can also combine with organic molecules such as ethyl or methyl groups.

The acidity of the environment, as measured by pH, is a major factor in determining the valence state of a metal. Metals are oxidized at low pH, that is, they move from a low ionization state to a higher ionization state (e.g., elemental Zn or Zn^0 to Zn^{2+}). These ionic forms readily form reactions with oxygen to form oxides (such as rust) or sulfur to form sulfides. Metals must be in ionic form before they can be assimilated; therefore, ionic metals in the water are more biologically active than elemental forms.

Organometals (Figure 4.2) can be intentionally produced for a variety of industrial or agricultural purposes, but they can also occur naturally. Most often the organic portions augment the toxicity of the metal itself because they facilitate biological assimilation and ease the passage into cells. Inside the cells the organometals can become DNA adducts and cause cancers, malformations, and other toxic effects. Some of the metals used to form organometals include mercury, boron, silicon, selenium, germanium, tin, lead, arsenic, and platinum. Some naturally occurring organometals include selenomethionine and methylmercury, and these are well known for producing serious environmental effects.

In the 1980s, a great deal of concern was given to acid precipitation, which is produced when sulfates (SO_4^{-2}) or nitrogen oxides (NO_2^-, NO_3^-) are released through the burning of fossil fuels such as coal or oil. These by-products are transported through the atmosphere and precipitate as wet (e.g., rain) or dry (particulate) deposition. Upon entering lakes or streams, these pollutants reduce the pH of water and

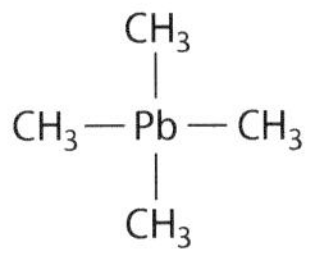

Tetraethyl lead—used to be in leaded gasoline

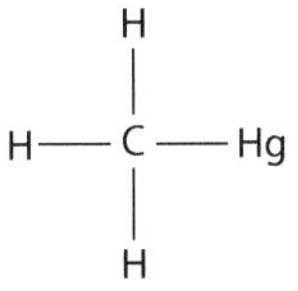

Methylmercury—occurs naturally in wetlands

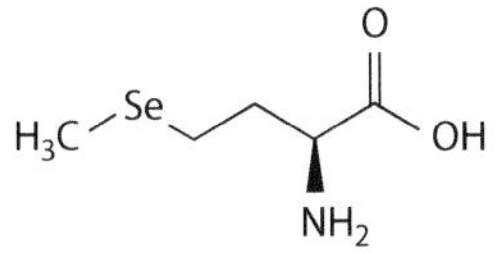

Selenomethionine—a naturally occurring organic form of selenium that is also an essential amino acid

FIGURE 4.2 Some common forms of organometals.

increase the solubility of metals. Aluminum (Al) has been closely associated with the toxic effects of acid precipitation. Aluminum is the most common metal and the third most common mineral in the earth's crust, so it is everywhere. The valence state of Al varies from 2^- to 3^+, with cations in acidic systems. In acidic conditions, dissolved Al has resulted in reduced reproduction among several plants, invertebrates, fish, and amphibians. We found that acidification can also affect the growth and development of young waterfowl because aluminum forms insoluble complexes with phosphorus, preventing the normal development of bony tissue (Sparling 1991). Acid precipitation came under intense scrutiny in the 1980s when the United States and Canada became concerned about increased acidity and its effects, especially in the northeastern United States and eastern Canada. Acid deposition is also a problem in northern Europe.

Metals tend to bind with organic matter either in water or soil. This is primarily a physical binding and need not involve actual chemical reactions. Binding makes metals less bioavailable than free, dissolved metals in the water column. **Dissolved organic carbon** (DOM) is composed of organic molecules that are dissolved in bodies of water such as lakes and rivers. As the concentration of DOM increases, the availability of contaminants that are attracted to the DOM decreases to aquatic organisms. Similarly, soils and sediments with high organic content are generally less toxic than those with low content.

Calcium and other metals affect metal toxicity in another way. Calcium in soil or water often exists as calcium carbonate ($CaCO_3$). Calcium carbonate buffers acidity, thus reducing the solubility of metals. In addition, the calcium ion (Ca^{2+}) is preferentially taken up by the living cells compared to many metals. Thus, in environments with moderate to high calcium, soluble concentrations of metals may be reduced and that which is in solution is inhibited from being assimilated, thus further reducing toxicity. Metals with biological functions such as zinc may also preferentially bind to cells and inhibit the uptake of other metals. This is called **competitive binding** and it can ameliorate the effects of more toxic metals.

4.4   BIOLOGICAL EFFECTS OF METALS

When Paracelsus (refer to Chapter 1) stated that the "dose makes the poison," it is quite possible that metals or metalloids like arsenic were on his mind. Metals such as iron, copper, and zinc are needed by virtually all plants and animals. In addition, humans require cobalt, molybdenum, selenium, manganese, and magnesium. Some plants also need cobalt, boron, and nickel. There is a balance in all organisms between sufficient intake of metals and toxic intake (Figure 4.3). At very low concentrations physiological problems can occur because the processes dependent on the metals cannot take place. At somewhat higher levels minor physiological problems that might be difficult to diagnose can occur. At even higher concentrations, mild toxicity might begin to appear, and at even higher intake, serious toxicity can occur. Somewhere in between deficiency and toxicity is the optimal range for a particular species.

This dynamic means that even biologically essential metals can exert a wide variety of toxic effects given sufficient doses. Those that are not essential may have no

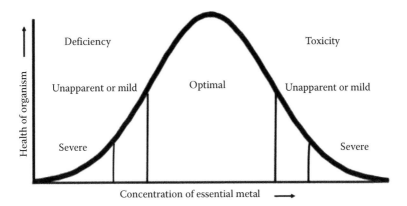

FIGURE 4.3 The relationship between exposure to essential metals and toxicity. At low exposure concentrations, deficiencies may result in health problems. High concentrations may cause toxicity, but in between those two extremes is an optimal range that provides sufficient metal without causing health issues.

effect at low concentrations but eventually they will exert toxicity. As we discussed in Chapter 3, metals can chelate with enzymes, thereby interfering with the very processes that low doses of these enzymes synergize. Depending on the metal and exposure concentration, serious neurological, metabolic, reproductive, teratogenic, and immunological damage can occur through exposure. For instance, both lead and Hg, two of the most toxic metals, can produce very serious neurological effects in humans and wildlife.

There is a built-in protective system against metal toxicity in many organisms. **Metallothionein** (MT) is a family of cysteine-rich, low molecular weight proteins that have the capacity to bind metals, whether they are essential or not. Metallothioneins are found across the fungi, plant, and animal kingdoms and are even found in some bacteria. Metallothioneins provide protection against metal toxicity and oxidative stress and are involved in the regulation of essential metals. Their production is induced by the presence of metals and other minerals in the bloodstream. Metallothioneins may bind selectively with a wide range of metals including cadmium, zinc, mercury, copper, arsenic, and silver. Thionein, the organic basis for metallothionein, picks up a metal when it enters a cell and carries the metal to another part of the cell where it is released or secreted. Cysteine residues from MTs can capture harmful oxidant radicals like the superoxide and hydroxyl radicals. In this reaction, cysteine is oxidized to cystine, and the metal ions that were bound to cysteine are released to the bloodstream and away from cells.

Although plants tend to be more tolerant to metal exposure than animals, they too can experience problems with photosynthesis, discoloration of leaves, reduced growth, and root death. In contrast, some plants are **hyperaccumulators** and can uptake, translocate, and tolerate high levels of some heavy metals that would be toxic to most other organisms. Studies have shown that these various plants can have leaves with more than 100 mg/kg of Cd, 1,000 mg/kg of Ni and Cu, or 10,000 mg/kg of Zn and Mn (dry weight) when grown in metal-rich environments (Kamal et al. 2004).

It is often difficult to precisely define the toxicity of a metal under natural conditions; the mere presence of a metal, even at concentrations that have been reported as toxic, does not mean that toxicity will occur. For instance, the valence state (elemental or some ionic form) may result in very different properties. Some metals may assume a variety of valence states and, given all other things equal, demonstrate a range of toxicity. Chromium usually occurs in its elemental state, as Cr^{3+} ion, or as Cr^{6+} ion. The Cr^{6+} ion is more toxic than the other two forms. Greater toxicity for Cr^{6+} has been observed for a wide range of organisms from yeast (*Saccharomyces cerevisiae*, Huang et al. 2014) to cancer in rats (Collins et al. 2010).

Trying to identify specific LC50 or LD50 data for a given animal species and metal is fraught with difficulties. In addition to the environmental factors listed above, life stage, age, sex, and reproductive condition of the animal all affect toxicity. Also, different laboratories use different methods of exposure—some use injection, others gavage, while some put the metal in the diet or drinking water. Some dissolve the metal in acidic water and use that, others use metals combined with other molecules such as acetate to aide in the solubility of the metal. Some expose the organism for 24, 48, or 96 hours or set up a chronic test. As a result, the best we can hope for is a range of values to estimate toxicity for a given species, and it is very important to understand how a toxicity test was conducted. Despite these variables, we can safely make the generalization that most metals are toxic to aquatic animals in the µg/L concentrations and in the mg/kg range for birds and mammals. It is also safe to say that sublethal effects occur at much lower concentrations than acute toxicity and pose the biggest problems for free-ranging organisms. Copper and zinc differ from most other metals in that, globally, their environmental deficiencies are more of a concern than their toxicity (Eisler 2000).

4.5   CHARACTERISTICS OF LEAD AND MERCURY

In this section we will discuss in greater detail two of the most toxic metals—lead and mercury. These two metals differ in several physical properties, but they share common characteristics of potentially producing serious sublethal effects and even death at comparatively low concentrations.

The U.S. EPA (2014) lists several metals as **priority pollutants**, so named because they are commonly found in sewage effluents and have defined protocols for their analysis. This list for metals includes antimony, arsenic, beryllium, cadmium, chromium, copper, lead, mercury, nickel, selenium, silver, thallium, and zinc. These are not the only metals that can exert toxicity, as we have suggested above, other metals can be toxic at high concentrations and may have environmental relevance in highly contaminated sites.

4.5.1   CHARACTERISTICS OF LEAD

In 2016 the Centers for Disease Control and Prevention (CDC) stated that "today at least 4 million households have children living in them that are being exposed to high levels of lead. There are approximately half a million U.S. children ages 1–5 with blood lead levels above 5 micrograms per deciliter (μg/dL), the reference level at which CDC recommends public health actions be initiated. No safe blood lead level in children has been identified." Shortly before this chapter was written, officials in the city of Flint, Michigan, which has a large proportion of its population living below the poverty line, became greatly concerned for, after a switch in water sources, its citizens as they began to have measurable concentrations of lead in their drinking water and were strongly advised to install water filters in their homes. Subsequent investigations in other cities revealed that a significant proportion of Americans have lead in their drinking water. According to the World Health Organization (WHO 2015), childhood lead exposure is estimated to contribute to about 600,000 new cases of children developing intellectual disabilities and 143,000 deaths every year.

Lead (Pb) is soft, malleable, and heavy; in fact it is the heaviest nonradioactive element. It has an atomic number of 82 and an atomic mass of 207.2 g. In nature it occurs most commonly as Pb^0 or Pb^{2+}. Most often lead occurs in various oxides and sulfides.

About half of the amount of Pb mined each year goes into leaded batteries (ATSDR 2007). Other common uses include small arms ammunition, shotgun pellets, fishing sinkers, and weights for fishing nets and tire balancing. Less common uses include electrodes, radiation shielding, building industry, sculptures, leaded glass, ceramic glazes, semiconductors, and some industrial paint pigments, although its use in house paint was banned in 1978 when it was found that children who ate leaded paint suffered many serious illnesses. However, lead paint still occurs in poorly maintained slums and houses and is a source of concern. Lead was used to seal plumbing joints until 1998 when it was banned by the EPA for that purpose. Organic lead, especially tetraethyl lead, was used in gasoline for automobiles from the 1920s until the early 1970s as an octane booster. After the organic portion of the molecule was burned away during combustion, the inorganic lead was emitted through tail pipes and entered the atmosphere or was deposited on roads where it could be washed into rivers and streams. Following EPA's ban, the output of Pb from vehicles declined by 95% and atmospheric Pb declined by 94% (EPA 2016a, Figure 4.4).

Nriagu (1978) estimated that the terrestrial environment contains about 99.9% of the total amount of lead in the world. That would put it at about 16 mg/kg in soils and 27 mg/kg in sediments (Eisler 2000). Eisler (2000) also reported that air typically has a Pb concentration of around 0.1 μg/m³ (1 part in 10^{-10}) to maybe 100 times that in heavy metropolitan areas. In industrial settings, mean concentrations of lead in air can exceed 5400 ug/m³. In comparison, the average concentration of lead in natural waters ranges from 0.02 μg/L in oceans to 2 μg/L in lakes and rivers and 20 μg/L in groundwater (Eisler 2000).

Acid water and dissolved salt concentrations heavily influence the amount of soluble Pb in surface waters. Acidity increases the solubility of lead, but hard water reduces its bioavailability. Sulfate ions, if present in soft water, limit the lead concentration in solution through the formation of lead sulfate. Most of the lead in water will be tied to sediments.

On land, lead strongly adheres to soil particles, which reduces the bioavailability of the metal to plants. Like in water, bioavailability increases as pH and soil organic

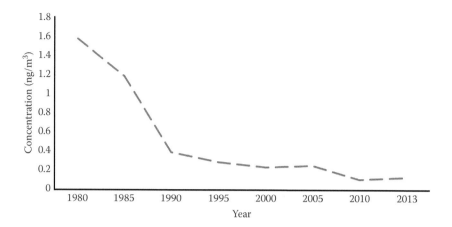

FIGURE 4.4   Decline in average atmospheric lead concentrations from 1980 to 2013. (Data from U.S. Environmental Protection Agency (EPA), List of priority pollutants, 2014, https://www.epa.gov/sites/production/files/2015-09/documents/priority-pollutant-list-epa. pdf, accessed November 22, 2016; U.S. Environmental Protection Agency (EPA), Lead, 2016a, http://www.epa.gov/airtrends/lead.html, accessed November 22, 2016.)

matter decrease. Most plants, if they assimilate Pb to any degree, seem to sequester it in their roots, allowing little to enter shoots or leaves. While there are literally hundreds of plant species that are known hyperaccumulators of metals, there are only a few that have been associated with lead, meaning that they can have over 1000 mg Pb/kg in their tissues (Auguy et al. 2013). Among the best known hyperaccumulators of lead are the leadworts, family Plumbaginaceae.

In uncontaminated sites, plants and invertebrates tend to have higher concentrations of the metal than birds or mammals. Aquatic invertebrates had 5–58 mg Pb/kg dry weight in their tissues, but some in contaminated sites have over 14,000 mg/kg (Eisler 2000).

Fish tend to have relatively low concentrations of Pb in their tissues. In a fairly extensive review, Eisler (2000) found that values were often <1 mg/kg live weight in clean areas but could get as high as 29 mg/kg in contaminated sites. For birds the range was from <0.1 mg/kg to around 8 mg/kg for the whole body. In all vertebrates, lead accumulates in bone, so we would expect to find higher concentrations there than in other tissues. Mammals are comparable to birds.

Grillitsch and Schiesari (2010) produced an extensive list of metal concentrations in reptiles. Concentrations in unpolluted sites were similar or lower than those observed in birds and mammals, while contaminated sites, of course, had higher concentrations. Among hundreds of data points, some high concentrations included 115 mg/kg wet weight in bones of snapping turtles (*Chelydra serpentina*), 136 mg/kg dry weight in the bones of box turtles (*Terrepene carolina*), and 105–386 mg/kg wet weight in captive alligators (*Alligator mississippiensis*) in Louisiana.

Terrestrial vertebrates assimilate Pb through inhalation or ingestion. Inhaled Pb enters the lungs and almost immediately enters the bloodstream, whereas ingested lead may be dissolved by acidic digestive systems and absorbed more slowly into the blood. Much of the ingested lead will be depurated with feces. High fiber diets or those with adequate calcium or iron will inhibit lead absorption.

As mentioned, lead is stored in bone, but that lead is essentially unavailable to the rest of the organism. Liver is probably the organ next in line for lead concentrations (Eisler 2000), and that lead is readily available for metabolism and toxicity. Organic forms of Pb are usually highest in kidneys and fat.

Lead can negatively impact every organism and virtually every biological system within organisms; it is a highly toxic, cumulative, metabolic poison. Lead can cause mutations, teratogenesis, and cancer; disrupt reproduction; impair liver and thyroid functions; and attack the immune system, but its primary target is the nervous system. One of the main ways that lead works is by binding with or deactivating many proteins and enzymes in plants and in animals.

When Pb is assimilated by plants, it can reduce photosynthesis, interfere with mitosis, inhibit growth, affect pollen germination and seed viability, and impair water absorption. Naturally, higher concentrations of lead in soil will result in higher plant concentrations, all other factors being the same.

For most aquatic biota, Eisler (2000) concluded that (1) dissolved waterborne Pb was the most toxic form, (2) organic lead compounds were more toxic than inorganic compounds, (3) adverse effects in some species were observed at 1 μg Pb/L, and (4) effects were more apparent at elevated water temperatures, reduced pH, comparatively soft waters, and in younger life stages.

A huge environmental problem in the past has been birds' intoxication by ingesting lead shot pellets. Hunting of birds such as waterfowl, doves, and others is an important sport in the United States and elsewhere, but the problem we are talking about is not due to the direct killing of these birds. Rather, if a duck or goose is shot but not killed and the shot imbeds in its tissues, the shot will be surrounded by a cyst that isolates it from the bird's physiology. However, if the bird ingests lead shot, severe problems may occur. The acidity of the bird's gut dissolves the lead and allows it to be absorbed by the bloodstream. Once this occurs, the many adverse effects of lead can be produced and the bird may slowly die. Prior to the 1990s, Friend (1989) and others estimated that 1.8–2.4 million waterfowl were dying each year due to ingested shot. Federal legislation banning lead shot on National Wildlife Refuges occurred in 1991, and this was expanded to all waterfowl hunting and in wetlands owned by the federal government. Subsequently, many states have enacted laws restricting or banning the use of lead shot. More recently, states are looking at restricting lead shot from public areas used by dove hunters.

In addition to lead shot, lead fishing sinkers have resulted in waterfowl deaths as they can be mistaken for submerged seeds. Concern about lead weights began in England in the 1970s and 1980s and centered on mute swans (*Cygnus olor*) that were found dead or dying. Common loons (*Gavia immer*) are also at risk (Scheuhammer and Norris 1996). In 1987 Britain partially banned the use of leaded weights, and the swan population has increased in size since then, but it is not clear how much of the increase was due to the lead ban. Several countries have also either banned the use of lead in sinkers or are promoting voluntary discontinuation. Secondary poisoning caused by scavenging carcasses of animals that were killed by hunters and contain fragments of lead bullet or shot is also of concern, especially for the critically endangered California condor (*Gymnogyps californianus*) and other large raptors.

Several generalizations on mammalian toxicity have been developed from extensive research on the effects of lead in laboratory and wild animals: (1) lead toxicosis can occur under real environments with actual exposure concentrations; (2) organic lead is usually more toxic than inorganic lead; (3) there is considerable variation among species in sensitivity to Pb; and (4) as with assimilation of Pb, many environmental factors including calcium, magnesium, pH, and organic matter can affect lead toxicity; diets deficient in some basic nutrients such as calcium, minerals, and fats can contribute to lethal and sublethal expressions (Eisler 2000). Signs of Pb poisoning in mammals include those seen in birds but also involve spontaneous abortions, blindness, peripheral nerve disease, poor performance in tests involving learning or memory, and various blood disorders.

4.5.2 GENERAL CHARACTERISTICS OF MERCURY

Mercury (Hg) is a contaminant of global concern. Several international conferences, agreements, and conventions have been conducted to regulate and reduce Hg concentrations in the environment, particularly as they relate to human health. As we mentioned above, mercury is unique among metals in several ways. Mercury has no known biological function, it is the only metal that is liquid under standard temperature and pressure conditions, and it can form organic complexes such as methylmercury (MeHg), which are fat soluble and can both bioconcentrate and biomagnify through food chains. In addition, mercury is resistant to most acids, although concentrated sulfuric acid can dissolve the metal. In contrast, Hg dissolves several other metals to form amalgams. Gold and silver are two commercially useful amalgams with mercury. Dental amalgams using Hg are used to fill in tooth cavities, although their popularity is fading. Mercury is also infamous for forming an amalgam with aluminum by dissolving the lighter metal. For this reason, the transport of Hg by aircraft is largely banned—you can imagine what might happen if the container with Hg developed a leak or spilled!

Mercury is an exceptionally rare metal in the earth's crust; it comes out as the 66th most common element in the crust and has an average concentration of 0.08 mg/kg in soils and sediments (U.S. EPA 2016b). That means that only 0.000008% of the earth's soil is Hg. When found, however, Hg tends to pool and form relatively rich but widely scattered pockets. The most common valence states for mercury are 0, 1, and 2; higher valence states can occur but are very rare. The 1+ oxidation state often takes the form of Hg_2^{+2}, with two Hg atoms forming a dimer with a 2+ charge. All forms of inorganic Hg are toxic but elemental Hg is less toxic than its ionic forms. As we will see, Hg has a tendency to become organic by becoming methylated or ethylated under certain natural environmental conditions and the organomercury forms are far more toxic than inorganic Hg.

The most common Hg-containing ore is cinnabar (HgS). Around 1810 tonnes of Hg are extracted each year, with China accounting for 75% of the total production (USGS 2016). The last Hg mine in the United States closed down in 1992.

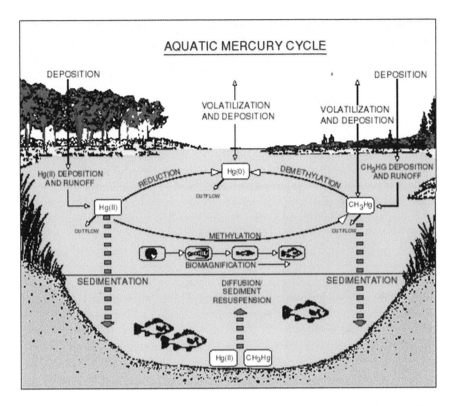

FIGURE 4.5 Mercury cycle. Mercury enters waterways through atmospheric deposition or runoff. In water, mercury can be methylated or it can enter sediments where it is at least partially removed from aquatic organisms. Mercury can volatize and reenter the atmosphere. During these events the ionic form of mercury can change. Mercury can also be assimilated by microorganisms and biomagnify. (Courtesy of the USGS, http://wi.water.usgs.gov/mercury/mercury-cycling.html.)

Mercury occurs in the lithosphere, atmosphere, hydrosphere, and biosphere. Since it occurs in both organic and inorganic forms in all of these media and since bacteria are involved in the conversion of Hg from one state to another, a **mercury cycle** has been recognized (Figure 4.5). Starting with the atmosphere, either inorganic or organic Hg can be transported long distances until it eventually precipitates on land or water. The primary sources for the 5500–8900 tonnes of Hg emitted into the atmosphere each year are natural processes such as volcanoes and combustion in coal-fired plants. Together these account for 82% of atmospheric Hg (Pacyna et al. 2006).

On land or water, Hg can volatilize as elemental Hg back into the atmosphere, enter the food chain, or run off into water bodies as one ionic form or elemental Hg. Mercury in runoff or effluents amasses to around 1000 tonnes per year globally (UNEP 2013). Much of the Hg that enters water bodies settles into sediments and can be locked in place for millions of years until some natural event such as a volcano releases it back into the atmosphere.

Mercury is persistent. While its mean retention time in the atmosphere is around 6–90 days, it can last in soils and sediments for essentially forever since it is an element and more than 2000 years in ocean water before it is eventually recycled into the atmosphere. Ocean sediments account for 330×10^9 tonnes, ocean water for 4.15×10^9 tonnes, soils for 21×10^6 tonnes, the atmosphere for 850 tonnes, and freshwater for only 4 tonnes (Eisler 2000). Although the atmosphere is not a primary reservoir for holding Hg, it serves as a major conduit from one reservoir to another.

Mercury has been used in industry, medicine, and agriculture, but its demand has about halved since 1980 primarily due to the concern for its toxicity (Wilburn 2013). In industry, Hg has been used in measuring devices such as thermometers, in float valves, for electrical purposes, and in the production of some types of telescopes. Today, Hg in thermometers and other instruments has been greatly replaced with colored alcohol or digital devices. In the 1880s, Hg was used as a bath in making felt hats. The metal neatly separated the fur from the skin to obtain felt. This was in addition to the arsenic that they were also using to preserve the skins. Unfortunately, hat manufacturers did not recognize the neurotoxicity produced by Hg, and many hat makers came down with neurological disorders. This is where the saying "mad as a hatter" and the rather weird Mad Hatter in Lewis Carroll's novel *Alice's Adventures in Wonderland* came from.

In agriculture, Hg has been used as a seed coating to protect against fungal and bacteria. Unfortunately, some people ate seed corn and other seeds protected with the organomercury coating and became poisoned. This practice of coating seeds with Hg was stopped in the United States in 1995 but still occurs in some other nations.

In medicine, the use of Hg goes all the way back to ancient Egyptians and China. It is still used in Chinese medicine but rarely in Western medicine. As mentioned above, Hg is used with other metals in dental amalgams. It is also found in thimerosal as a preservative for vaccines and related fluids.

Organomercury and inorganic Hg are very different with regard to bioaccumulation. Both can bioconcentrate but organomercury can easily do it, and only organomercury biomagnifies as one progresses up an aquatic food chain. Organomercury is more completely absorbed or assimilated from foods than is inorganic mercury; it is also more soluble in lipids, passes more readily through biological membranes, and is slower to be excreted (Eisler 2000). All of these factors add up to considerably more risk to organisms from organomercury than from the inorganic forms of the metal.

Among plants in clean freshwater, the average concentration of Hg is around 0.10 mg/kg live weight, but this can increase to 10 or 11 mg/kg at contaminated sites (Eisler 2000). Freshwater invertebrates have baseline concentrations around 0.20 mg/kg live weight in clean and 3–4 mg/kg in contaminated sites. Marine invertebrates averaged 0.40 mg/kg in reference and 6–7 mg/kg in polluted sites.

We humans generally eat the muscle of fish and therefore are most interested in how much Hg is present in that tissue. For lean fish, mean Hg concentrations in muscle or whole bodies seldom exceed 0.5 mg/kg. Nicole Greenfield of the Natural Resources Defense Council (2016) lists the following fish in their hazardous category (I have added mean Hg concentrations in mg/kg based on FDA (2014) data), which should be entirely avoided: king mackerel (0.485), orange roughy (0.871), marlin (0.485), shark (0.979), swordfish (0.995), tilefish (1.45), and bigeye or ahi tuna (0.689). Other species have recommendations for limited consumption. The article has a longer list of relatively safe seafood, including anchovies (great on pizza), catfish, clams, oysters, crab, and trout. Stahl et al. (2009) conducted an EPA-sponsored study of 500 lakes across the country looking for Hg and other contaminants in predaceous and bottom-dwelling freshwater fish. They found Hg in all of their fish samples, with methylmercury accounting for around 90% of all mercury detected. Mercury concentrations in total bodies of predaceous fish ranged from 0.023 to 6.605 mg/kg fresh weight with a median value of 0.28 mg/kg. Bottom-dwelling fish had values ranging from 0.05 to 0.596 mg/kg with a median value of 0.069 mg/kg. The authors mentioned that mercury accounts for about 90% of the fish consumption advisories in the United States. Because organisms are quick to take up Hg and slow to lose it, older fish, as measured by age, length, or weight, tend to have higher concentrations than young ones from the same waters.

The 96 hr LC50 for *Daphnia* sp. is 5 µg/L (EPA 1980), and this copepod is a model for several other species of freshwater invertebrates. Lethal concentrations of total Hg generally range from 0.1 to 2.0 µg/L for sensitive fish, but the high end may be around 150 µg/L to produce a 96 hr LC50. Among the sensitive species of fish, adverse sublethal effects can occur at water concentrations of 0.03–0.1 µg Hg/L.

Among amphibians and reptiles, Hg concentrations generally ranged from below detection limits to <0.5 mg/kg fresh weight (Eisler 2000). However, amphibians collected below a synthetic fiber mill had Hg concentrations that were 3.5–22 times those of the same species collected from reference streams; mean values were 3.45 mg/kg for adults and 2.48 mg/kg for subadults (Bergeron et al. 2010).

Considerable research has been conducted on Hg levels in birds and mammals. Most tissues and whole bodies have <1.0 mg/kg, but aquatic birds and mammals tend to have higher concentrations than terrestrial species. High concentrations in birds from contaminated sites can exceed 100, even 200 mg/kg. Carnivorous birds generally have higher concentrations of Hg than grain or plant eaters. Feathers can be a reliable, nondestructive source to determine if a bird has picked up Hg, and the truly neat thing is that they can be taken from museum specimens for a historic perspective. Hair offers a similar benefit for mammals.

Marine mammals often have higher mercury, especially organomercury, concentrations than terrestrial species. That is partially because marine mammals tend to occupy trophic levels next to top carnivores and are subject to biomagnification. Mercury concentrations in marine mammals typically measure below 15 mg/kg (Eisler 2000), but values reaching into several hundred mg/kg have been recorded in seals. In contrast, elevated concentrations in nonmarine mammals averaged around 5 mg/kg or lower and seldom exceeded 50 mg/kg (Eisler 2000).

Birds and mammals are substantially more resistant to Hg toxicosis than are most aquatic animals. Median lethal doses of organomercury are in the neighborhood of 2.2–50 mg/kg body weight or 4.0–40 mg/kg in the diet among sensitive birds. Toxicity of inorganic mercury is much, much lower. For instance, in Japanese quail (*Coturnix japonica*), the LC50 using inorganic Hg was between 2956 and 5086 mg/kg in the diet over 5 days of exposure, followed by 7 days of observation. However, only 31–47 mg/kg dietary methylmercury was needed to produce 50% lethality. Given as an acute oral dose, 26–54 mg/kg inorganic Hg and 11–33.7 methylmercury produced LC50s over a 14-day observation period in the same species (Hill and Soares 1987).

Among mammals, a 96 hr LD50 of 17.9 mg/kg methylmercury given orally was obtained for mule deer (*Odocoileus hemionus*, Hudson et al. 1984). Mink are particularly sensitive to Hg and it only took 1.0 mg/kg to produce 100% mortality of the test animals over a 2-month period (Sheffy and St. Amant 1982).

Sublethal effects in birds and mammals include teratogenesis, mutagenesis, and cancer. It can negatively affect reproduction and growth, behavior, and blood chemistry. Nervous system effects can include loss of motor coordination, hearing, vision, and (in humans) thought processes. Tissue damage in kidneys, livers, pancreas, and heart are common. For sensitive birds, sublethal toxicity can occur at 640 µg Hg/kg body weight or 50–500 µg/kg in the diet. Sensitive mammals may be affected at 250 µg Hg daily/kg body weight or 1100 µg/kg in the diet.

We cannot leave our discussion about Hg without mentioning a major human tragedy due to methylmercury. From 1932 through 1968 and beyond, the Japanese-owned Chisso Corporation dumped untreated effluents into the Minamata Bay, Japan. In 1968 the company installed some minor treatment of effluents after the government began to complain, but it ignored warning signs of an impending disaster. The source of Hg was a chemical production plant. The resulting effluents were also heavily contaminated with several other metals and other pollutants. Through this period, shellfish in the bay bioaccumulated exceedingly high concentrations of methylmercury. The shellfish served as a major portion of the diets of these people. The first human victim was a young girl in April 1956, who experienced convulsions and problems with walking. By May of that year, an epidemic of neurological conditions was evident. Patients with the '**Minamata disease**' suffered a loss of sensation and numbness in their hands and feet, problems with running or walking, hearing, seeing, and swallowing. Even domestic cats and fish-eating wildlife were observed having spasms and convulsions. By November 1956, investigators identified that the one common factor among humans and animals was a diet high in fish and shellfish. They concluded that a heavy metal was involved in the disease and Chisso became the chief suspect. Due to obstruction and subterfuge by the company, there was a long delay in determining that organic mercury was the specific toxic agent. By March 2001, 2,265 victims had been officially certified as having Hg poisoning (1,784 of whom have died) and over 10,000 people had received some government-mandated financial compensation from Chisso. As a result of this disaster, the United Nations established the Minamata Convention on Mercury in 2013. The principal articles of this international convention include a ban on new Hg mines, the phaseout of existing ones, control measures on air emissions, and regulation of small-scale gold mining (UNEP 2015).

4.6 CHAPTER SUMMARY

1. Metals occupy most of the Periodic Chart of Elements. With the exception of mercury, they are malleable, ductile, fusible, opaque, solid at room temperature, good conductors of heat and electricity, and are easily ionized; typically contain two to three electrons in their outermost valence shell; and have luster.
2. Some metals are essential nutrients, which results in a dynamic ranging from possible deficiency to toxicity.
3. Heavy metals that include lead, mercury, chromium, copper, nickel, zinc, and cadmium tend to be of greater ecotoxicological concern than other metals, although all can be toxic given sufficient concentrations.
4. Factors that affect metal toxicity include the acidity of the medium, organic content including dissolved organic carbons, water hardness, and the presence of certain other metals that may compete for binding or uptake.
5. Some metals readily combine with organic molecules, and these organometals tend to be more easily assimilated and are more toxic than inorganic forms of the same metal.
6. Lead and mercury are very toxic to humans, animals, and plants. Sources of lead in the environment include mining, smelting, and industrial applications that are not tightly regulated. Lead toxicity is expressed in many ways including teratogenic malformations, mutations, cancer, and kidney and liver diseases. Bone serves as a significant repository for this metal.
7. Mercury also produces many harmful effects, but it is mainly a neurotoxin.

4.7 SELF-TEST

1. Which of the following characteristics of metals refers to their ability to melt and blend with other metals?
   a. Malleable
   b. Fusible
   c. Ductile
   d. None of the above
2. Based on its behavior at room temperature, which is the most unique of metals?
   a. Lead
   b. Chromium
   c. Selenium
   d. Mercury
3. What is a good definition of a heavy metal as they pertain to ecotoxicology?
   a. Metals that are denser and have a molecular weight greater than iron
   b. Metals that are found in the environment
   c. Any metal that is toxic
   d. An older type of music played by Led Zeppelin and Deep Purple
4. What did Paracelsus mean when he said that "the dose makes the poison"?

5. When contaminants are ingested and are incorporated into cells they are said to be
   a. Bioaccumulated
   b. Assimilated
   c. Bioconcentrated
   d. Depurated
6. What is a major way for metals that are located deep in the earth to be recycled back into the surface?
   a. Volcanoes
   b. Upheavals such as mountain formation
   c. Tsunamis
   d. More than one above
7. Describe the relationship between deficiency and toxicity for those metals that are essential for diets.
8. True or False. Inorganic mercury can become organomercury under natural conditions—that is, without human involvement.
9. True or False. The concentration of lead in the atmosphere of the United States has increased in recent years due to an increase in industrial pollution.
10. Briefly describe why ingested lead is more toxic to waterfowl than are lead pellets that have become imbedded under a duck's skin.

REFERENCES

Agency for Toxic Substances and Disease Registry (ATSDR). 2007. *Toxicological Profile for Lead*. Atlanta, GA: Agency for Toxic Substances and Disease Registry.

Auguy, F., Fahr, M., Moulin, P., Brugel, A., Laplaze, L. et al. May 7, 2013. Lead tolerance and accumulation in *Hirschfeldia incana*, a Mediterranean Brassicaceae from metalliferous mine spoils. *PLoS One*. DOI: http://dx.doi.org/10.1371/journal.pone.0061932.

Bergeron, C.M., Bodinof, C.M., Unrine, J.M., Hopkins, W.A. 2010. Mercury accumulation along a contamination gradient and nondestructive indices of bioaccumulation in amphibians. *Environ Toxicol Chem* 29: 980–988.

Centers for Disease Control. 2016. Lead. http://www.cdc.gov/nceh/lead/. Accessed November 22, 2016.

Collins, B.J., Stout, M.D., Levine, K.E., Kissling, G.E., Melnick, R.L. et al. 2010. Exposure to hexavalent chromium resulted in significantly higher tissue chromium burden compared with trivalent chromium following similar oral doses to male F344/n rats and female B6C3F1 mice. *Toxicol Sci* 1198: 368–379.

Eisler, R. 2000. *Chemical Risk Assessment; Health Hazards to Humans, Plants, and Animals*, Vol. 1, Metals. Boca Raton, FL: Lewis Publishers.

Friend, M. 1989. Lead poisoning: The invisible disease. U.S. Fish and Wildlife Leaflet 13.2.6. https://www.nwrc.usgs.gov/wdb/pub/wmh/13_2_6.pdf. Accessed March 9, 2017.

Grillitsch, B., Schiesari, L. 2010. Metal concentrations in reptiles. An appendix of data compiled from the existing literature, pp. 553–901. In: Sparling, D.W., Linder, G., Bishop, C.A., Krest, S.K. (eds.), *Ecotoxicology of Amphibians and Reptiles*, 2nd Edition. Boca Raton, FL: SETAC/CRC Press.

Hill, E.F., Soares, J.H. 1987. Oral and intramuscular toxicity of inorganic and organic mercury-chloride to growing quail. *J Toxicol Environ Health* 20: 105–116.

Huang, Z.W., Kuang, X., Chen, Z.X., Fang, Z.J., Wang, S., Shi, P. 2014. Comparative studies of tri- and hexavalent chromium cytotoxicity and their effects on oxidative state of *Saccharomyces cerevisiae* cells. *Curr Microbiol* 69: 448–456.

Hudson, R.H., Tucker, R.K., Haegle, M.A. 1984. *Handbook of Toxicity of Pesticides to Wildlife*. Resource Publication 153. Washington, DC: U.S. Fish and Wildlife Service.

Kamal, M., Ghaly, A.E., Mahmoud, N., Côté, R. 2004. Phytoaccumulation of heavy metals by aquatic plants. *Environ Int* 29: 1027–1039.

Natural Resources Defense Council. 2016. Mercury guide. https://www.nrdc.org/stories/mercury-guide. Accessed December 3, 2016.

Nriagu, J.O. (ed.). 1978. *The Biogeochemistry of Lead in the Environment. Part A. Ecological Cycles*. Amsterdam, the Netherlands: Elsevier/Holland Biomedical Press.

Pacyna, E.G., Pacyna, J.M., Joze, M., Steenhuisen, F., Wilson, S. 2006. Global anthropogenic mercury emission inventory for 2000. *Atmos Environ* 40: 4048–4063.

Scheuhammer, A.M., Norris, S.L. 1996. The ecotoxicology of lead shot and lead fishing weights. *Ecotoxicology* 5: 279–295.

Sheffy, T.B., St. Amant, J.R. 1982. Mercury burdens in furbearers in Wisconsin. *J Wildl Manage* 46: 1117–1120.

Sparling, D.W. 1991. Acid precipitation and food quality: Effect of dietary Al, Ca and P on bone and liver characteristics in American black ducks and mallards. *Arch Environ Contam Toxicol* 21: 281–288.

Stahl, L.L., Snyder, B.D., Olsen, A.R., Pitt, J.L. 2009. Contaminants in fish tissue from US lakes and reservoirs: A national probabilistic study. *Environ Monit Assess* 150: 3–19.

United Nations Environmental Programme (UNEP). 2013. Global mercury assessment 2013: Sources, emission, releases and environmental transport. Geneva, Switzerland: UNEP Chemical Branch, pp. 299–305.

United Nations Environmental Programme (UNEP). 2015. Minamata convention on mercury. http://www.mercuryconvention.org/Convention/tabid/3426/Default.aspx. Accessed November 22, 2016.

U.S. Environmental Protection Agency (EPA). 1980. Ambient water quality criteria document for mercury. EPA 440/5-80-058. Washington, DC: Office of Water.

U.S. Environmental Protection Agency (EPA). 2014. List of priority pollutants. https://www.epa.gov/sites/production/files/2015-09/documents/priority-pollutant-list-epa.pdf. Accessed November 22, 2016.

U.S. Environmental Protection Agency (EPA). 2016a. Lead. http://www.epa.gov/airtrends/lead.html. Accessed November 22, 2016.

U.S. Environmental Protection Agency (EPA). 2016b. The toxic release inventory (TRI) program. http://www2.epa.gov/toxics-release-inventory-tri-program. Accessed November 22, 2016.

U.S. Food and Drug Administration (FDA). 2014. Mercury levels in commercial fish and shellfish. http://www.fda.gov/Food/FoodborneIllnessContaminants/Metals/ucm115644.htm. Accessed December 3, 2016.

U.S. Geological Survey (USGS). 2016. Mercury statistics and information. http://minerals.usgs.gov/minerals/pubs/commodity/mercury/. Accessed November 22, 2016.

Wilburn, D.R. 2013. Changing patterns in the use, recycling, and material substitution of mercury in the United States. U.S. Geological Survey Scientific Investigations Report 2013-5137. http://pubs.usgs.gov/sir/2013/5137/. Accessed November 22, 2016.

World Health Organization (WHO). 2015. Lead poisoning and health. http://www.who.int/mediacentre/factsheets/fs379/en/. Accessed November 22, 2016.

5 Current Use Pesticides

5.1 INTRODUCTION

Pesticides have had a long history, extending to ancient times. Sumerians (5000–2000 BCE) used sulfur compounds to control weeds. The Chinese also used sulfur and later mercury and arsenic compounds to reduce body lice and household insect pests. This 'cure' may have been worse than the problem due to the toxicity of mercury and arsenic. Ancient Romans also used sulfur to control insects and salt to kill weeds. However, pest control, even up to the nineteenth century, was mostly conducted manually by having slaves pick off insects or manually pulling weeds.

The use of chemicals to control insects really caught on after World War II, with the introduction of organochlorines (OCs). The first and foremost of these was dichlorodiphenyltrichloroethane (DDT). Soon, other OCs including lindane, aldrin, dieldrin, endrin, toxaphene, and several others appeared. These pesticides were easily applied on crop fields, relatively inexpensive, and initially effective in controlling insect pests on a variety of plant crops. However, after several years, problems began to appear with these OCs. For one thing, they were very persistent with half-lives measured in decades. Also, insects began to develop a tolerance to DDT, which forced farmers to use more and more of the pesticide to get the same effect. Third, because of their ability to biomagnify, concentrations of these pesticides accumulated in higher trophic levels like fish-eating birds, which eventually caused population declines (see Chapter 6).

Starting in 1972 with DDT, the U.S. Environmental Protection Agency (EPA) began banning OC pesticides. This process took some time, however, with the most recent organochlorine to be banned, endosulfan, being phased out in the United States starting in 2010. However, some less persistent OCs are still in use for specific purposes.

As OCs were being phased out, new pesticides came on the market. These 'new generation' pesticides included organophosphates (OPs) and carbamates, which were first discovered in the 1930s and 1940s, but were not extensively used until OCs were on their way out. A little later, pyrethroids were synthesized based on a natural insecticide derived from the chrysanthemum flower *Chrysanthemum cinerariifolium* or *C. coccineum*. All three of these groups are still in use today and account for a large share of pesticides applied to crops and for household use. The science of pesticide development has grown tremendously over the past couple of decades, with new chemicals and even entire families of chemicals being synthesized. Active ingredient formulations are developed to control weeds, mites, nematodes, fungi, and many other pests. Half-lives of these new pesticides are typically measured in days, weeks, or a few months rather than years or decades.

5.2 WHAT IS A PESTICIDE?

By the roots of the word, a 'pesti-cide' is, obviously, something that kills pests. In this broad definition, a pesticide could include mechanical killing, burning, biological control, chemical death, or any other method of permanently removing a pest. But in this chapter, we will focus on manufactured chemical control agents that are currently being produced and are registered for use by the U.S. EPA; we'll also briefly discuss some inorganic and biological agents.

5.2.1 PESTICIDE USE IS CONTROVERSIAL

Not everyone agrees on the use of pesticides. On one hand, some argue that all pesticides are dangerous and that we are poisoning our streams and rivers with pesticide runoff from agricultural fields. They also argue that most pesticides are not species specific and that many beneficial insects and plants are victims to spraying. To some extent these are valid concerns, especially when they deal with misuse of pesticides.

Proponents of chemical pesticides, however, are also correct when they state that agriculture in its current form within North America, consisting of hundreds to thousands of acres of a single crop (Figure 5.1), would not be sustainable without pesticides. They point out that only with the aid of pesticides can we produce sufficient food for the world's human population. Without pesticides, the production of farm commodities would suffer and the global problems of famine would be far worse.

FIGURE 5.1 Modern agricultural practices rely on fields that are hundreds of acres in size and planted with a single crop, such as this field of rye. Monoagriculture of this type requires pesticides to be productive. (Courtesy of Pixaby 2017, https://pixabay.com/en/village-field-view-field-crops-264776/.)

The bottom line, of course, is that pesticides are neither bad nor good in and of themselves. It is how humans use pesticides that can be the problem. Indiscriminate spraying can cause substantial damage to the environment, so the use of pesticides should follow carefully prescribed instructions to minimize the damage they can cause. In the United States, the **Federal Insecticide, Fungicide, and Rodenticide Act** (FIFRA 1947; EPA, 2017), substantially amended in 1972, is the chief federal law affecting the registration and use of pesticides. All commercially manufactured pesticides must be registered with the EPA and follow usage regulations certified by the FIFRA. Registration requires that the manufacturers provide documentation on the research they have conducted to assess the toxicity of the compound to a variety of plant and animal models. Also, manufacturers must provide detailed directions on how the pesticide will be applied, such as what active ingredient the pesticide will be mixed with, where it will be used, when it will be used, methods of application, and application concentrations. While the process of registration is usually adequate, there are concerns in that only a select handful of species are approved by the EPA for testing, and these are not necessarily the ones that will be exposed to the chemical; endangered species are a special example of this (Racke and McGaughey, 2012). Also, for the most part, the EPA allows manufacturers to do their own testing and report the results rather than having objective, independent labs do the toxicity evaluations (Boone et al., 2014). This suggests a possible opportunity for conflict of interest to occur.

5.3   ECONOMICS OF CURRENT USE PESTICIDES

The pesticide industry is economically huge. For starters, the economic savings due to crop protection and disease reduction are formidable. Also, according to an industry-sponsored agency, the cost to put a new pesticide on the market takes an average of 9 years and $152–$256 million (McDougall, 2016). Third, government involvement with registration and enforcement of laws is extensive, with municipal, county, state, and federal agencies participating. The EPA is the chief federal agency overseeing pesticides, and we can get some idea of the cost to regulate these chemicals by looking at EPA budget requests, which is $679 million in 2017 or 8.3% of the agency's total budget request (EPA, 2016c). Annually, pesticide sales within the United States accounted for $11.78 billion in 2006 and $12.45 billion in 2007. Globally, sales amounted to over $35 billion in both years (EPA, 2016a).

The cost of the ecological effects of pesticides is the flip side of the coin, so to speak. In addition to crop protection and other benefits from pesticides, there is also a significant detrimental component. Pimental (2005) estimated that the annual major economic and environmental losses due to the application of pesticides in the United States were

- Public health issues—$1.1 billion
- Increased pesticide resistance in pests—$1.5 billion
- Crop losses caused by improper application of pesticides (e.g., herbicides)—$1.4 billion
- Bird deaths due to pesticides—$2.2 billion
- Groundwater contamination—$2.0 billion

Benefits from the use of pesticides include

- Increased food production
- Lower cost of food
- Better quality of foods
- Control of household pests
- Protection of industry and infrastructure from damage that would have been caused by blocked waterways, obstructed highways and reduced access to highway, utility and railroad rights of way
- Improved recreation areas, including lawns, gardens, public parks, athletic fields, lakes, and ponds
- Reduced threats to human, livestock, and pet from diseases, allergies, parasites, and other factors

It might be possible with extensive research to derive price estimates for these benefits, but they certainly reach into billions of dollars each year.

5.4 TYPES OF PESTICIDES

There are several types of 'families' of pesticides. Some of these families contain scores of pesticides, others only a few. One way of classifying pesticides is by the type of active ingredient chemical that they contain. We will discuss some of these below. Another way of classifying pesticides is by the **target organism**, the group of plants or animals that the chemical and its application were designed to control. Plants as weeds and animals as crop depredators or disease vectors can be at serious economic and health risks. Some pesticides have broad action and can be toxic to both plants and animals; others are more specific for plants or animals and often even for subcategories within each. We can conveniently group pesticides into the following categories based on their target organisms (Table 5.1).

TABLE 5.1
Types of Pesticides and Their Target Organisms

Types of Pesticides	Target Organisms
Acaricide	Mites, ticks, spiders
Microbiological	Bacteria, viruses, other microbes
Avicide	Birds
Fungicide	Fungi
Herbicide	Plants
Insecticide	Insects
Molluscicide	Snails, slugs
Nematicide	Nematodes
Piscicide	Fish
Rodenticide	Rodents
Sanitizer	Microbes

There can be inadvertent effects from pesticides such as injuring other plants or animals, which are collectively called **nontarget organisms**. However, most of these side effects can be minimized by following specified application directions. For example, there are federal and state laws banning the use of pesticides over wetlands without specific permits. Applicators who fly over crop fields must close their release valves before flying over wetlands and must be aware of the prevailing winds to avoid drift over; otherwise, they can receive heavy fines.

The number of different kinds of pesticides is mind boggling. The Pesticide Action Network (2014) lists the following types of pesticides, followed by the number of chemical families for each: insecticides, 20 families; herbicides, 36; biological agents, 8; fungicides, 12; microbiological, 11; inorganic/metals, 16; soaps, solvents, and adjutants (surfactants, carriers, and such), 11; others, 7. Altogether, the EPA registers more than 1375 pesticides (Scorecard, 2011), and you can bet that there are many other unregistered methods of chemically controlling pests. In addition to the synthesized chemical pesticides this list includes, biologicals, organically derived products, and oils are often included as solvents or adjutants, but they may have biological properties. Below are brief descriptions of a few of the more common pesticide families; the list is not exhaustive.

5.4.1 CARBAMATES

A carbamate is an organic compound derived from carbamic acid (NH_2COOH, Figure 5.2a). Carbamates, along with a related group of pesticides called organophosphates, largely replaced the OCs in the 1970s and are very widely used today. Carbamates include herbicides, fungicides, and insecticides. If you are a home gardener, you may have used Sevin®, which contains carbaryl, one of the many carbamates. Carbaryl (Figure 5.2b) was introduced in 1956, and it has been used globally more than all other carbamates combined (Fishel, 2014). Carbaryl is deadly at low concentrations to insects but has low toxicity to birds and mammals. A few other carbamates include aldicarb, metam sodium, and mancozeb. Literally, thousands of tons of these pesticides are sold each year.

Lower levels of carbamate toxicity can cause muscle weakness, dizziness, sweating, and slight body discomfort. At higher doses, headache, excess salivation,

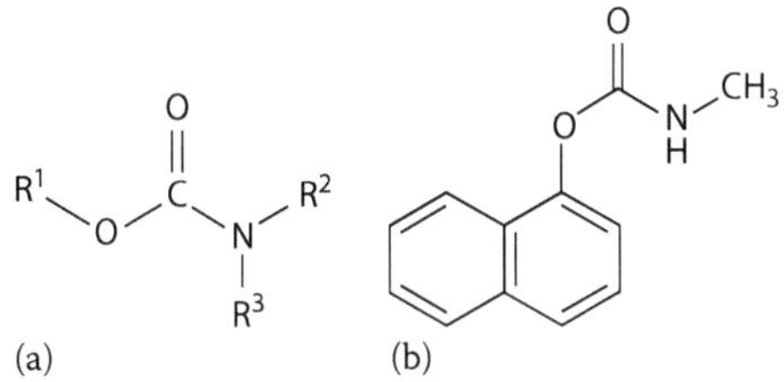

FIGURE 5.2 Carbamates, a commonly used group of pesticides: (a) a generalized carbamate. Each R refers to a different molecule or element but all carbamates have carbamic acid moiety with an oxygen doubly bonded to a carbon, which is connected to a nitrogen; (b) carbaryl, a common carbamate used in households.

nausea, vomiting, abdominal pain, and diarrhea can occur. In animals, carbamates function by inhibiting acetylcholinesterase (AChE), as we mentioned in Chapter 3. Carbamates only temporarily bind with the enzyme and release it after a period of time. Therefore, animals that survive the initial effects of the contaminant are likely to recuperate. In plants, carbamates interfere with the functioning of other enzymes.

Among the chemicals in this family, the most toxic include aldicarb and carbofuran. With both chemicals, the manufacturers, after receiving notice that the EPA was considering banning the products, voluntarily requested that they be at least partially deregistered, meaning that they could not be produced or used in the United States. They are also prohibited from use in Canada, the European Union, and elsewhere.

What is arguably the worst pesticide-related industrial accident ever occurred was in Bhopal, India, in 1984. A massive leak consisting of 30 tonne of methyl isocyanate and other gases occurred at a plant owned by Union Carbide India Limited. Methyl isocyanate is an intermediate chemical in the manufacturing of carbamates and is extremely toxic and irritating. During the night, the gas blew into the shanty towns surrounding the pesticide plant, gassing thousands of people. Officially, the initial death toll was nearly 3000, but even after three decades, the actual numbers are poorly estimated and have exceeded 4000. In addition, it is estimated that more than 16,000 people died prematurely due to the lingering effects of the leak and hundreds were blinded (Malik, 2014).

5.4.2 ORGANOPHOSPHATES

Organophosphates became popular in the mid-1970s along with carbamates. Their most common use is as insecticides, but some are registered as herbicides or fungicides. Like carbamates, OP insecticides are neurotoxins that bind with the enzyme AChE and prevent the breakdown of acetylcholine in the synapse between neurons and the neuromuscular junction. They are similar in structure and function to nerve gases such as sarin. All OPs have a phosphorous atom doubly bonded to either an oxygen or a sulfur atom and singly bonded to three other oxygen atoms (Figure 5.3). A few extensively used OP insecticides include parathion, malathion, methyl parathion, chlorpyrifos, diazinon, dichlorvos, phosmet, and fenitrothion.

OP pesticides are generally regarded as relatively safe for use on crops and animals due, in part, to their fast degradation rates compared to most OC pesticides. Their half-lives vary as a function of microbial composition, pH, temperature, and availability of sunlight. Depending on conditions, OPs usually have half-lives from a few days to a few months. OPs have relatively low log K_{ow} values (2–3) and thus have moderate to high solubility (10–10,000 mg/L).

Many different plants and animals metabolize OPs, so bioconcentration can occur but biomagnification does not. Some OPs exert their toxic effects directly without chemical transformation. Others, however, must be oxidized or sulfonated to a different chemical condition called oxon or sulfone before they have maximum toxicity. This oxidation can occur in the environment or through metabolism in animals by the P450 system. Chlorpyrifos, diazinon, and malathion are examples of OPs that are converted to oxon forms. The conversion makes the insecticides more efficient in binding with AChE and may increase their toxicity by 10–100 times, at least in

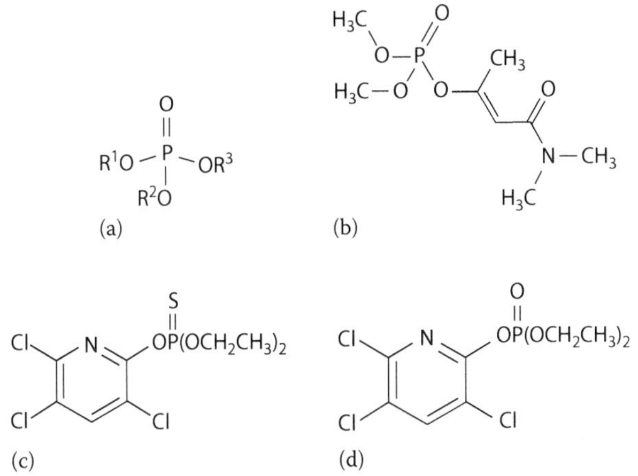

FIGURE 5.3 Organophosphate (OP) pesticides: (a) a generalized organophosphorus pesticide. While the constituents at the R sites might differ, all OPs have a phosphorous atom doubly bonded to either an oxygen or a sulfur atom and singly bonded to three other oxygen atoms. (b) Dicrotophos, an OP insecticide. (c) Chlorpyrifos, the most widely used organophosphate insecticide; note the doubly bonded sulfur instead of oxygen. (d) is the oxon or metabolized form of chlorpyrifos, which is many times more toxic than the parent chemical.

amphibians (Sparling and Fellers, 2007). These same pesticides have been linked to amphibian declines in pristine mountainous areas in California (Sparling and Fellers, 2009; Sparling et al., 2015).

While very toxic to insects, most OPs have moderate to low toxicity to birds and mammals. Table 5.2 presents some lethal dose concentrations of OPs to these groups. Note the high exceptions to mammalian toxicity for azinphos-methyl, methyl parathion, and phorate. In the United States, azinphos-methyl was phased out by the EPA, with the last permissible use on certain crops in 2012; it has also been banned in other nations. As with the other contaminants, aquatic organisms such as fish are more sensitive to OPs than mammals and birds. Bees and other pollinating insects can also be very sensitive to certain OPs.

5.4.3 PYRETHROIDS

The chrysanthemum plant (Figure 5.4) is native to Eastern Europe and Asia. It produces a natural insecticide, **pyrethrum**, that has been used for thousands of years as a mosquito repellent and as a remedy for lice in ancient Persia and elsewhere. Pyrethrum is actually a natural mixture of six chemicals called **pyrethrins** (ATSDR, 2003a, Figure 5.5a). You can make that same insecticide at home if you are growing chrysanthemums. Just dry the flowers, remove and crush the seeds, and rinse them in a small amount of water, and you have an effective insecticide that you can apply on your skin to keep mosquitoes away.

There have been at least two generations of **synthetic pyrethroids** based on the original pyrethrum. Pesticides in both generations are still used, although the

TABLE 5.2

Some Representative Acute Toxicity Values of Organophosphorus Pesticides to Rats (Oral), Rabbits (Dermal), Birds, Fish, and Bees

Pesticide	Rat Oral LD50 (mg/kg Body Weight)	Rabbit Dermal LD50 (mg/kg Body Weight)	Bird Acute Oral LD50[a]	Fish Acute in Water LC50[b]	Bee[c]
Azinphos-methyl	4	150–200	HT	VHT	HT
Chlorpyrifos	96–270	2000	HT	HT	HT
Diazinon	1250	2020	HT	HT	HT
Disulfoton	2–12	3.6–15.9	HT	MT	HT
Malathion	5500	>2000	MT	HT	HT
Methyl parathion	6	45	NA[d]	ST	NA
Naled	191	360	MT	HT	HT
Phorate	2–4	20–30 (guinea pig)	VHT	ST	MT
Phosmet	147–316	>4640	ST	HT	HT

Source: Fischel, F. M., *Pesticide Toxicity Profile: Carbamate Pesticides*, Institute of Food and Agricultural Sciences, University of Florida, Gainesvilla, FL, 2014. http://edis.ifas.ufl.edu/pi088, accessed November 22, 2016.

[a] *Bird toxicity LC50 (mg/kg body weight)*: ST (somewhat toxic) = 501–2000; MT (moderately toxic) = 51–500; HT (highly toxic) = 10–50; VHT (highly toxic) ≤ 10.

[b] *Fish toxicity LC50 (mg/L)*: ST = 10–100; MT = 1–10; HT = 0.1–1; VHT ≤ 0.1.

[c] Bee toxicity MT = moderately toxic (kills if applied on bees); HT = highly toxic (kills upon contact as well as residues).

[d] Data not available.

FIGURE 5.4 *Chrysanthemum coccineum*, one of the two species that provide natural pyrethrum.

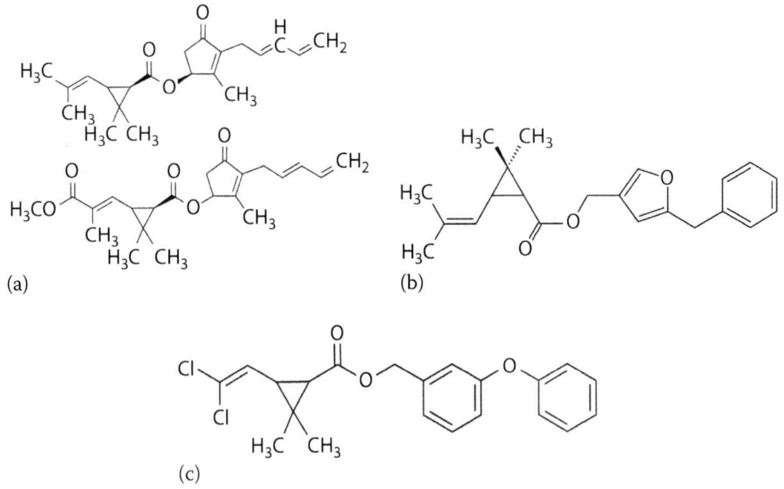

FIGURE 5.5 Pyrethrins and pyrethroids. (a) Two or six pyrethrins or natural pyrethroids. (b) Resmethrin, a first-generation pyrethroid. (c) Permethrin, a second-generation synthetic pyrethroid. Note the chlorine atoms that contribute to the stability of this compound.

second generation ones have larger sales volumes. The first-generation pyrethroids, developed in the 1960s, include bioallethrin, tetramethrin, resmethrin (Figure 5.5b), and bioresmethrin. They are more efficient than natural pyrethrins but are not stable in sunlight. Thus, they have very short half-lives. The second-generation pyrethroids, such as permethrin (Figure 5.5c), cypermethrin, and deltamethrin, were introduced in the mid-1970s and are significantly more stable than those of the first generation. However, these second-generation chemicals are more toxic to mammals than the earlier formulations.

Sales of pyrethroids are less than the popular carbamates or OPs, in part, due to the higher insect toxicity of pyrethroids. That may sound counterintuitive, but when insecticides are very effective in insect control, there is less need for large quantities. In addition, pyrethroids seem to be used more in home products than in agriculture but are in over 3500 registered products (EPA, 2016b). For instance, allethrin is the active ingredient in Raid® insect spray, bifenthrin is an active ingredient in Ortho Home Defense Max®, and cyfluthrin and other pyrethroids are used in Baygon®, another home insecticide. Flea and tick powders for dogs and cats typically contain first-generation pyrethroids that are effective against ticks and safe for pets. Caution should be used with cats, however, for they do not have one of the enzymes necessary for the rapid breakdown of pyrethroids and could become sick when cleaning their fur.

These chemicals have very low vapor pressure (10^{-5} to 10^{-7} mm), which equates to low volatility and, with the exception of two or three, they are nearly insoluble in water. Accordingly, their log K_{ow} values are around 5–8 (ATSDR, 2003a).

Even the second-generation pyrethroids tend to be very short-lived. Half-lives are measured in hours or days in air or water and up to a few weeks in soils and sediments. Because they have low vapor pressures, air pollution is less of a concern than with other insecticides. Also, because they are generally insoluble, pyrethroids will not mix in the water column of lakes or streams, but tend to strongly adhere to dissolved organic carbons, other suspended particulates, and sediments where bio-availability to free-swimming organisms is low. Even still, under natural conditions, pyrethroids can be highly toxic to aquatic insects and moderately high to highly toxic to fish and amphibians. Median lethal concentrations range from 1 to 20 µg/L and these concentrations are experienced under natural conditions.

Pyrethroids are also very toxic to terrestrial insects. Fortunately, birds and mammals can metabolize pyrethroids quickly, and the metabolism does not make the pesticides more toxic. Thus, lethal concentrations for birds and mammals exceed expected environmental levels, and sublethal toxicity tends to be unimportant at environmentally realistic concentrations (ATSDR, 2003a).

Pyrethroids are neurotoxins whose outward signs may be similar to organophosphate or carbamate exposure, but the mechanisms are different. While they do affect what goes on in synapses and inhibit AChE, pyrethroids primarily react along nerve axons and extend firing of the neurons. Outward signs of toxic exposures include loss of coordination, convulsions, and paralysis (Soderlund and Bloomquist, 1989). Like OPs and carbamates, death is often due to respiratory failure.

The most common symptom of pyrethroid exposure in humans is a temporary numbing of the skin called **paresthesia**. Pyrethrins have been classified as possible human carcinogens because long-term studies on rats resulted in elevated frequencies of liver cancer (ATSDR, 2003a). Neither pyrethroids nor pyrethrins have been shown to produce genotoxic, endocrine, or reproductive effects in mammals. Mild immunological suppression occurred in rats and mice with permethrin, deltamethrin, and cypermethrin, usually following weeks of exposure in food (ATSDR, 2003a).

5.4.4 Phosphonoglycine

There are only two chemicals of concern in this family of pesticides—glyphosate (Figure 5.6a) and glufosinate-ammonium. But glyphosate sales are huge, over $6 billion per year. Corn and soybeans are the major crops for glyphosate. In 1970, Monsanto developed the herbicide glyphosate, which has become the most widely used herbicide in North America. Glyphosate tops the EPA's list of the 25 most widely used pesticides. It is also used in many other nations, especially in Asia-Pacific, China and Japan.

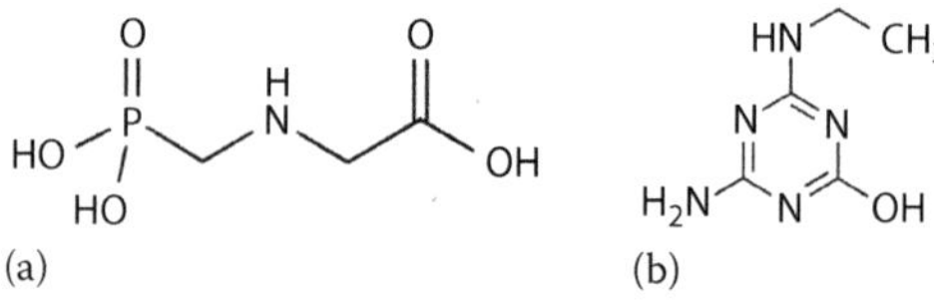

FIGURE 5.6 Phosphonoglycine glyphosate and triazine atrazine: (a) Glyphosate, the most widely used herbicide on the market today. Note that carbon atoms exist at the crooks of the structural formula. (b) Atrazine, the second most widely used herbicide.

Glyphosate is best known commercially as Roundup®. One reason that Roundup sales are huge is Monsanto has also been working with genetic modification to make crop plants more resistant to the herbicide. Roundup is used as a post-emergent herbicide, meaning that it is applied after weeds (and often crops) have germinated through the soil. In the past, glyphosate could not be used once crops emerged because they would succumb to the herbicide just like weeds. By genetically making the crops resistant to the herbicide, farmers can now apply the herbicide, get good control of weeds, and not worry about crop mortality. Crops that have been genetically modified so far include corn, canola, sugar beets, cotton, and soybeans. Specifically, glyphosate interferes with enzymes essential in the production of chlorophyll and hence interferes with photosynthesis.

Glufosinate is marketed as its ammonium salt, glufosinate-ammonium. It too is a post-emergent herbicide. Glufosinate is derived from a natural compound isolated from two species of *Streptomyces* fungi (as in streptomycin). Bayer's Crop Science has also started genetically modifying some crops such as corn, soybeans, and cotton to be resistant to this herbicide. Glufosinate inhibits the activity of an enzyme that is necessary for the breakdown of ammonia, allowing for toxic buildup of ammonia in cells. Glufosinate can also inhibit the same enzyme in animals.

Both chemicals have similar properties, so we'll just discuss glyphosate. They have high propensity to particulates in soil and are highly water soluble. Glyphosate readily and completely biodegrades in soil, even under low temperature conditions. Its average half-life in soil is about 60 days but may last longer in cold environments. In field studies of cold regions, residues may be found the following year and may have residual effects. If glyphosate enters aquatic systems through accidental spraying, spray drift, or surface runoff, it will dissipate rapidly due to absorption into dissolved organic matter and biodegradation. Its half-life in water is usually a few days. Glyphosate does not bioconcentrate to any extent. The herbicide has a very low volatility, so atmospheric concentrations are usually negligible.

Field studies suggest that the chemicals associated with glyphosate may be more toxic than the herbicide itself. The formulation approved for control of aquatic vegetation, Rodeo®, seems to be relatively safe to aquatic animals. However, Roundup has been accidentally or illegally sprayed on wetlands, causing die-offs of fish, amphibians, and aquatic invertebrates. An early study (Relyea, 2005) showed that Roundup was lethal to both larvae (i.e., tadpoles) and juveniles of three species of frogs at environmentally relevant concentrations. Subsequently, Relyea and Jones (2009) determined that the 96 hour LC50 for nine species of larval frogs using Roundup Original ranged from 2.2 to 5.6 mg active ingredient (a.i.)/L, while that for three species of larval salamanders, it ranged from 7.6 to 8.9 mg (a.i.)/L. A typical in-field concentration of glyphosate is around 4.2 mg/L, about midway of the LC50s in frogs, but lower than that for salamanders. Recall, however, that longer exposures require lower concentrations to have similar effects, so chronic exposures at these concentrations are probably lethal under natural conditions. Relyea's team of students have also found that the timing and quantity of exposures can influence the extent of mortality and that glyphosate can also alter competition among tadpoles and exert other complex effects.

Technical grade glyphosate has very low acute toxicity to most terrestrial animals. Oral LD50 values for glyphosate are greater than 10,000 mg/kg in mice, rabbits, and goats. It is practically nontoxic to skin exposure and is not even irritating to the skin of rabbits or guinea pigs. Glyphosate is slightly toxic to wild birds. The dietary LC50 in both mallards and bobwhite quail is greater than 4500 mg/kg (EXTOXNET, 1996). It is also practically nontoxic to most fish, although some species are sensitive, but it may be slightly toxic to aquatic invertebrates.

5.4.5 TRIAZINES

Triazines consist of a benzene or phenol ring with three nitrogen atoms substituting for three carbons (Figure 5.6b); the potential hydrogen atom sites can also be substituted with other molecules to increase the diversity of these compounds. Triazines were first developed in the early 1950s. All triazines are synthetic. As an interesting side note, one of the triazines is the basis for melamine, a resin that has been used for making Melmac® dishware and other products, which were widely popular in the 1940s through the 1960s. In this section, however, we will focus on only one triazine, the herbicide atrazine.

Atrazine is the second most widely used herbicide in the United States, only being surpassed by glyphosate (EPA, 2016a). Between 33,000 and 35,000 tonnes of atrazine is applied in the United States each year. Due to environmental and human health concerns, however, atrazine was banned by the European Union in 2003. Atrazine is a selective herbicide used to prevent pre- and post-emergent broadleaf weeds in crops. It is very widely used on corn and on turf such as golf courses and residential lawns. Atrazine has a relatively high solubility of 33 mg/L. Compared with other triazines, atrazine is a solid at room temperature and has low volatility, making it rare in the atmosphere.

Atrazine is moderately persistent, with half-life measured a few weeks to several months. However, carryover of atrazine and other triazine residues through the winter has been known to damage plants in the year following application. Typical half-life in surface water is around 200 days and about 14 days in air. Photooxidation is not important in breaking down these chemicals in the atmosphere or in surface waters, but in air, ionic oxidations reduce its half-life by several hours (ATSDR, 2003b).

Acute toxicity from atrazine does not pose a particularly high risk to birds and mammals. In rats, for instance, the oral LD50s (96 hour) are from 1900 to 3000 mg/kg body weight, depending on how it's given to the animals. In rabbits, LD50 through injection is 750 mg/kg. An example of lethal dietary toxicity in mallards was 19,650 mg/kg in food during an 8-day exposure and 5,600 mg/kg in northern bobwhite (Eisler, 2000).

As with many contaminants, acute toxicity of triazines in general and atrazine specifically is higher in aquatic invertebrates than in mammals and birds and run in the low mg/L region. For example, for the water flea, *Daphnia magna*, the 48 hour LC50 for atrazine was 6.9 mg/L, 5.7 mg/L for the scud (a species of fish) *Gammarus fasciatus*, and 0.7 mg/L for the midge *Chironomus tentans* (Macek et al., 1976). LC50s (96 hour) for rainbow trout (*Oncorhynchus mykiss*), bluegill (*Lepomis macrochirus*), and zebrafish (*Danio rerio*) were 4.5–24, 8–42, and 37 mg/L, respectively (reviewed by Eisler, 2000).

While triazines are not as acutely toxic as some of the other pesticides we have discussed, continuing research is revealing that atrazine may produce several sublethal effects. In the laboratory, atrazine targets the reproductive system of developing mammals. It disrupts estrus (the reproductive cycle of most female mammals, except humans), affects hormone concentrations in the blood, increases the frequency of spontaneous abortions and miscarriages, decreases the body weights of fetuses, and results in poor bone formation in fetal rats (ATSDR, 2003b). Many of the studies reported these effects at atrazine concentrations much higher than we would ever expect in field studies, but again, longer-term exposures at lower concentrations may produce similar results. Among humans that have been exposed to atrazine in the field or industry, there is a statistically significant shortened pregnancy, an uptick in the rate of prostatitis (inflammation of the prostate gland), and some embryonic malformations (ATSDR, 2003b).

Atrazine has been involved in a heated scientific debate for the past several years regarding amphibians. Bishop et al. (2010) provided an excellent summary of the effects of atrazine on amphibians and reptiles and on the controversy. Between 1980 and mid-2016, over 100 studies appeared in the peer-reviewed literature focusing on amphibians and atrazine—an impressive showing for a narrowly focused topic. Although atrazine is not persistent, its extensive use makes it frequently available to amphibians at various life stages. Giddings et al. (2005) found that atrazine is found primarily in the water column and not adsorbed to sediments, thereby increasing the risk of exposure to aquatic life stages of amphibians. Very high environmental concentrations of the herbicide in water are around 200–300 µg/L, but effective concentrations are orders of magnitude less. In their review of atrazine and amphibians, Bishop et al. (2010) found that there is much greater risk from sublethal effects under typical environmental conditions than outright mortality.

Two main camps about atrazine toxicity and amphibians developed in the early part of this century and both were prolific in publishing scientific papers, so we cannot review them in their entirety. Suffice to say that the principal initiator in favor of a negative atrazine effect was Tyrone Hayes from the University of California, Berkeley. Among other things, Dr. Hayes published studies that showed that atrazine can feminize male frogs, meaning that instead of a gonad that displays only male characteristics males exposed to very low concentrations of atrazine developed gonads that contained both testes and ovaries (Hayes et al., 2002, 2006). Other investigators have supported that atrazine can induce a variety of harmful effects at environmentally relevant concentrations. On the other side were studies published by Dr. James Carr et al. (2003), Dr. Keith Solomon et al. (2008), and others, which seemed to show no negative effects at relevant concentrations. This has been one of the most contested ecotoxicological arguments in the past 15 years. Again, see Bishop et al. (2010) for a more complete review of this contentious issue.

5.4.6 Inorganics, Metals, and Biologics

There are hundreds of 'natural' compounds and elements that are used as pesticides, in addition to all of those that are synthesized in a factory. For example, there are over 50 chemicals that are classified as inorganics in the very comprehensive database supported by the state of California (California Department of Pesticide Regulation, 2016).

Leading the group in terms of sales are compounds based on sulfur, copper, bromine, chlorine, chlorates, boric acid, hydrogen peroxide, and zinc. From our short history of pesticides above, we saw that sulfur has been used for centuries and is still used today as a fungicide and insecticide. Copper compounds also have many target organisms among fungi and insects. Chlorine and chlorates are important fumigants to remove unwanted organisms from seeds prior to planting. Bromide is used in compounds primarily as a fungicide or microbiological agent. There are organic compounds such as ethyl or methyl alcohol that are used to sterilize seeds and equipment and petroleum distillates such as mineral spirits, kerosene, and naphtha that are used as pesticides in their own right or as solvents for other chemical agents. The risk of these compounds to nontarget organisms varies considerably from one product or organism to another, but, in general, this group has lower acute toxicity to vertebrates than many synthetic pesticide groups.

Biologics are products that are derived from or contain living organisms. Sales of plant-derived oils such as margosa or neem oil rank high in this category. Other oils that are used to repel insects include coconut oil, soybean oil esters, garlic, eucalyptus, rosemary, thyme, general vegetable oil, tall oil (a by-product of the Kraft process of coniferous tree pulp manufacturing), balsam fir, castor bean, cottonseed, black pepper, cedarwood, citronella, geranium, lemongrass, and peppermint. Some of these oils are lethal to insects or their eggs, while others coat fungal spores and insect eggs and suffocate them. Urea, a by-product of animal wastes, is used both as a high nitrogen fertilizer and in pre-emergent pesticides. The California database includes some unusual natural compounds such as sugar, coyote urine, fox urine, hydrolyzed corn, plant hormones, red cabbage color, yeast, buffalo gourd root powder, and putrescent whole egg solids.

One of the earliest and most widely used biological insecticides is the bacterium *Bacillus thuringiensis*, or *Bt*. This bacterium produces spores that contain proteins called δ-endotoxins. When insects ingest the spores, the spores break apart within the digestive tracts, the crystals become water soluble and release toxins that paralyze the digestive tracts of the insects, and the pests literally starve. Different strains of *Bt* are selective for specific insect larvae, including caterpillars, flies, mosquitoes, beetles, and nematodes. The California database also lists some strains of viruses that can infect caterpillars and fungi that attack beetle larvae or nematodes.

5.5   CHAPTER SUMMARY

1. Pesticides of various kinds are ancient, going back several thousand years, but the use of synthetic chemicals started in earnest in the 1940s.
2. Pesticide usage is controversial for they do contribute to environmental damage, especially if they are misused. However, we could not produce sufficient food to feed the world without the use of pesticides. Therefore, it behooves everyone to use pesticides wisely and according to prescribed directions.
3. Economics of pesticides include the cost of research and development, marketing and sales, environmental damage inflicted by pesticides, and governmental regulation of pesticide usage. Total costs associated with pesticides reach billions of dollars each year.

4. Pesticides may be classified by their chemical composition or by their target organisms.
5. Carbamates are common pesticides that work by inhibiting the enzyme acetylcholinesterase in the synapse of nerves, thereby triggering repetitive firing of neurons. They include insecticides, herbicides, and other types of pesticides. Perhaps, one of the greatest human-related disasters was caused by a leak of an intermediate chemical in manufacturing carbamates in Bhopal, India.
6. Organophosphates also inhibit AChE and tend to have short half-lives. The family includes some of the most widely used insecticides. Millions of tons of OPs are sold each year.
7. Pyrethroids are derived from the *Chrysanthemum* sp. flower but have been developed over the years into several synthetic chemicals. They are extensively found in products for control of insects around the home. They have very short half-lives, are essentially non-water-soluble, and do not volatilize; therefore they exert minimum environmental problems.
8. Phosphonoglycines include the herbicides glyphosate and glufosinate-ammonium. Glyphosate is the most widely used herbicide in the world. Sales of the product have increased geometrically after Monsanto, the manufacturer of Roundup, began genetically modifying crops to withstand the herbicide, thereby making it more effective as a post-emergent pesticide.
9. Triazines include atrazine, the second most widely used herbicide. Atrazine is effective as a pre- and post-emergent herbicide, but recent findings that it may cause health problems in humans have led to its banning in the European Union and elsewhere. Atrazine remains the focus of a debate concerning the welfare of amphibians.
10. Hundreds of inorganic compounds, biological organisms, and chemicals usable in organic farming have been registered by the U.S. EPA. Hundreds more undoubtedly are used without being registered. Among the inorganics are the standbys such as copper and sulfur that have been used since antiquity. Among the biologicals is *Bt*, a form of bacteria that can be used against specific insect pests.

5.6 SELF-TEST

1. What element has been used to fight crop pests since antiquity?
   a. Iron
   b. Zinc
   c. Sulfur
   d. DDT
2. Approximately when did organophosphate and carbamate pesticides come into the market?
   a. 1920s
   b. 1940s
   c. 1970s
   d. 1990s

3. What general type of pesticide has zebra mussels as its primary target?
   a. Miticide
   b. Molluscicide
   c. Rodenticide
   d. None of the above
4. Which of the following is a part of the total economic picture of pesticides?
   a. Cost of making regulations and enforcement
   b. Damage done by misuse or abuse of pesticides
   c. Research and development of new pesticides
   d. All of the above
5. True or False. The following diagram depicts a carbamate

6. True or False. Most insecticides are more toxic to terrestrial organisms than they are to aquatic organisms.
7. Which of the following is least likely to be found in the atmosphere or in the water column in substantial concentrations?
   a. Resmethrin
   b. Glyphosate
   c. Carbaryl
   d. Atrazine
8. What is the active ingredient in the herbicide Roundup?
   a. Atrazine
   b. Glufosinate
   c. Glyphosate
   d. None of the above
9. Short answer. What federal agency is most responsible for enforcing regulations concerning pesticides in the United States?
10. Short answer. Why is atrazine of concern to free-living amphibians?

REFERENCES

Agency for Toxic Substances and Disease Registry (ATSDR). 2003a. *Toxicological Profile for Pyrethrins and Pyrethroids*. Atlanta, GA: U.S. Department of Health and Human Services. http://www.atsdr.cdc.gov/toxprofiles/tp155.pdf. Accessed November 22, 2016.

Agency for Toxic Substances and Disease Registry (ATSDR). 2003b. *Toxicological Profile for Atrazine*. Atlanta, GA: U.S. Department of Health and Human Services. http://www.atsdr.cdc.gov/toxprofiles/tp153.pdf. Accessed November 22, 2016.

Bishop, C.A., McDaniel, T.V., de Solla, S. 2010. Atrazine in the environment and its implications for amphibians and reptiles. Pp. 225–260. In: Sparling, D.W., Linder, G., Bishop, C.A., Krest, S.K. (eds.), *Ecotoxicology of Amphibians and Reptiles*, 2nd Edition, SETAC. Pensacola, FL: CRC Press.

Boone, M.D., Bishop, C.A., Boswell, L.A., Brodman, R.D., Burger, J. et al. 2014. Pesticide regulation amid the influence of industry. *Bioscience* 64: 917–922.

California Department of Pesticide Regulation. 2016. Databases. http://www.cdpr.ca.gov/dprdatabase.htm. Accessed November 22, 2016.

Carr, J.A., Gentles, A., Smith, E.E., Goleman, W.L., Urquidi, L.J. et al. 2003. Response of larval *Xenopus laevis* to atrazine: Assessment of growth, metamorphosis, and gonadal and laryngeal morphology. *Environ Toxicol Chem* 22: 396–405.

Eisler, R. 2000. *Handbook of Chemical Risk Assessment: Health Hazards to Humans, Plants and Animals*, Vol. 2, Organics. Boca Raton, FL: Lewis.

EXTOXNET. 1996. Pesticide information profiles: Glyphosate. http://extoxnet.orst.edu/pips/simazine.htm. Accessed November 22, 2016.

Fischel, F.M. 2014. *Pesticide Toxicity Profile: Carbamate Pesticides*. Gainesvilla, FL: Institute of Food and Agricultural Sciences, University of Florida. http://edis.ifas.ufl.edu/pi088. Accessed November 22, 2016.

Giddings, J.M., Anderson, T.A., Hall Jr., L.W., Hosmer, A.J., Kendall, R.J. et al. 2005. *Atrazine in North American Surface Waters: A Probabilistic Aquatic Ecological Risk Assessment*. Pensacola, FL: Society for Environmental Toxicology and Chemistry.

Hayes, T.B., Collins, A., Lee, M., Mendoza, M., Noriega, N. et al. 2002. Hermaphroditic, demasculinized frogs after exposure to the herbicide atrazine at low, ecologically relevant doses. *Proc Natl Acad Sci USA* 99: 5476–5480.

Hayes, T.B., Stuart, A., Mendoza, M., Collins, A., Noriega, N. et al. 2006. Characterization of atrazine-induced gonadal malformations in African clawed frogs (*Xenopus laevis*) and comparisons with effects of an androgen antagonist (Cyproterone acetate) and exogenous estrogen (17β estradiol): Support for the demasculinization/feminization hypothesis. *Environ Health Perspect* 114(Suppl): 134–141.

Macek, K.J., Buxton, K.S., Sauter, S., Gnilka, S., Dean, J.W. 1976. Chronic toxicity of atrazine to selected aquatic invertebrates and fishes. U.S. EPA Rep 600/3-76-047. 58pp.

Malik, A. 2014. 30 Years after the Bhopal Disaster, India has not learned the lessons of the world's worst industrial tragedy. *International Business Times*. http://www.ibtimes.com/30-years-after-bhopal-disaster-india-has-not-learned-lessons-worlds-worst-industrial-1731816. Accessed November 22, 2016.

McDougall, P. 2016. The cost of new agrochemical product discovery, development and registration in 1995, 2000, 2005-8, and 2010 to 2014. https://croplife.org/wp-content/uploads/2016/04/Cost-of-CP-report-FINAL.pdf. Accessed December 4, 2016.

Pesticide Action Network (PAN). 2014. http://www.panna.org/.

Pimental, D. 2005. Environmental and economic costs of the application of pesticides particularly in the United States. *Environ Develop Sustain* 7: 229–252.

Racke, K.D., McGaughey, B.D. 2012. Pesticide regulation and endangered species: Moving from stalemate to solutions. Pp. 3–27. In: Racke, K.D., McGaughey, B.D., Cowles, J.L., Hall, A.T., Jackson, S.H. et al. (eds.), *Pesticide Regulation and the Endangered Species Act*. American Chemical Society Symposium Series, Vol. 1111, American Chemical Society, Washington, DC.

Relyea, R.A. 2005. The lethal impact of Roundup on aquatic and terrestrial amphibians. *Ecol Appl* 15: 1118–1124.

Relyea, R.A., Jones, D.K. 2009. The toxicity of Roundup Original Max to 13 species of larval amphibians. *Environ Toxicol Chem* 28: 2004–2008.

Scorecard. 2011. Federal regulatory program lists. http://scorecard.goodguide.com/chemical-groups/one-list.tcl?short_list_name=pest#top. Accessed November 22, 2016.

Soderlund, D.M., Bloomquist, J.R. 1989. Neurotoxic actions of pyrethroid insecticides. *Annu Rev Entomol* 34: 77–96.

Solomon, K.R., Carr, J.A., Du Preez, L.H., Giesy, J.P., Kendall, R.J. et al. 2008. Effects of atrazine on fish, amphibians, and aquatic reptiles: A critical review. *Crit Rev Toxicol* 38: 721–772.

Sparling, D.W., Bickham, J., Cowman, D., Fellers, G.M., Lacher, T., Matson, D.W., McConnnell, L. 2015. In situ effects of pesticides on amphibians in the Sierra Nevada. *Ecotoxicology* 24: 262–278.

Sparling, D.W., Fellers, G. 2007. Comparative toxicity of chlorpyrifos, diazinon, malathion and their oxon derivatives to larval *Rana boylii. Environ Poll* 147: 535–539.

Sparling, D.W., Fellers, G.M. 2009. Toxicity of two insecticides to California, USA, anurans and its relevance to declining amphibian populations. *Environ Toxicol Chem* 28: 1696–1703.

U.S. Environmental Protection Agency (EPA). 2016a. 2006–2007 pesticide market estimates: Sales. https://www.epa.gov/pesticides/pesticides-industry-sales-and-usage-2006-and-2007-market-estimates. Accessed November 22, 2016.

U.S. Environmental Protection Agency (EPA). 2016b. Pyrethroids and pyrethrins. https://www.epa.gov/pesticide-reevaluation/groups-pesticides-registration-review#pyrethroid. Accessed November 22, 2016.

U.S. Environmental Protection Agency (EPA). 2016c. FY2017 EPA budget in brief. https://www.epa.gov/sites/production/files/2016-02/documents/fy17-budget-in-brief.pdf. Accessed November 22, 2016.

U.S. Environmental Protection Agency (EPA). 2017. Federal Insecticide, Fungicide, and Rodenticide Act (FIFRA) and Federal Facilities. https://www.epa.gov/enforcement/federal-insecticide-fungicide-and-rodenticide-act-fifra-and-federal-facilities. Accessed April 21, 2017.

6 Halogenated Organic Contaminants

6.1 INTRODUCTION

This is a diverse group of chemicals collected here because they are organic, with carbon and hydrogen as principal constituents, but also contain some members of the group 17 elements from the Periodic Chart, halogens. Notably, the halogens are chlorine, bromine, and fluorine in order of importance; the other halogens are not typically present in contaminants. Most of the chemicals here are synthetic, having intentionally been designed for a particular purpose. They all tend to have long half-lives stretching into decades due to the strength of the carbon–halogen bond. Several have been banned as being persistent organic pollutants (POPs), and some are still in use. Most of these chemicals have two phenyl groups serving as their framework.

Families of compounds under consideration here include the following:

Polychlorinated biphenyls (PCBs)—A group of 209 compounds primarily used in electrical industries and banned in 1972 but still around in the environment due to their persistence (Figure 6.1a).

Note the single carbon–carbon bond for PCBs, the double oxygens in dioxins, and the single oxygen/carbon bond in furans.

Dioxins—More formally known as **polychlorinated dibenzo-p-dioxins (PCDD)**, these chemicals are by-products of burning organic molecules at high temperatures and can occur naturally or through industrial processes. Volcanoes and forest fires are two naturally occurring sources; incinerators are common industrial sources. They have no commercial purpose (Figure 6.1b).

Furans—Aka **polychlorinated dibenzofurans (PCDFs)**, these too are products of high combustion of organic molecules. Although they are not manufactured intentionally, except for research purposes, they may occur through incineration of refuse (Figure 6.1c) and can occur naturally.

Polybrominated diphenyl ethers (PBDEs)—PBDEs are currently used as flame retardants in clothing and other products due to their high combustion temperatures (Figure 6.2a).

Note the single oxygen bond for PBDE and the carbon–carbon bond for PBB.

Polybrominated biphenyls (PBBs)—Structurally identical to PCBs, except that bromine substitutes for chlorine, PBBs were phased out in the 1970s after a major contamination resulted in tens of thousands of domestic livestock being euthanized (Figure 6.2b).

Perfluorocarbons (PFCs)—Perfluorocarbons consist of alkane carbon chains and fluorine, and that's it—no hydrogens, nitrogens, or other elements

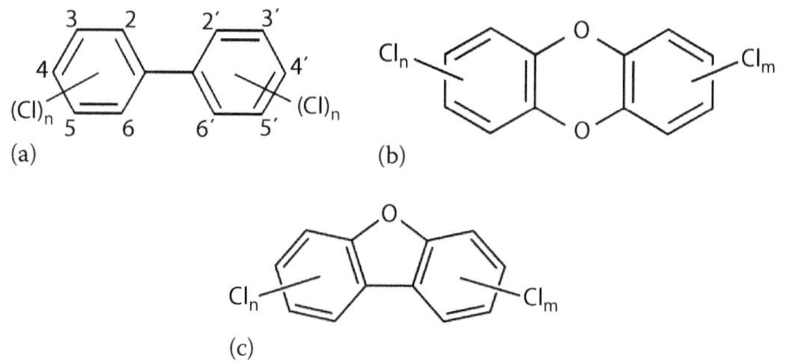

FIGURE 6.1 Generalized polychlorinated organic compounds: (a) generalized PCB; (b) generalized dioxin; and (c) generalized furan.

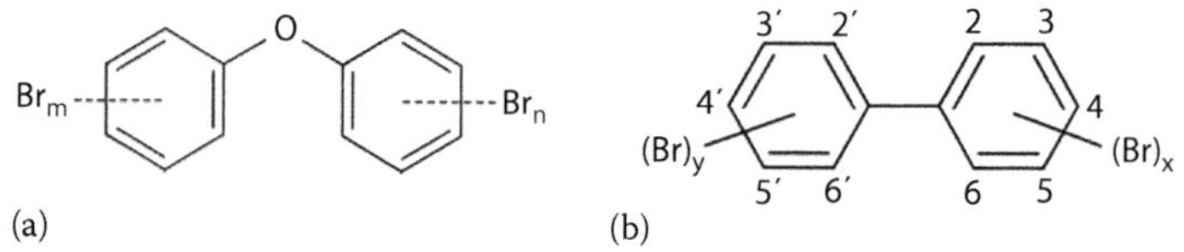

FIGURE 6.2 Generalized polybrominated molecules of environmental concern: (a) generalized polybrominated diphenyl ether; (b) generalized polybrominated biphenyl.

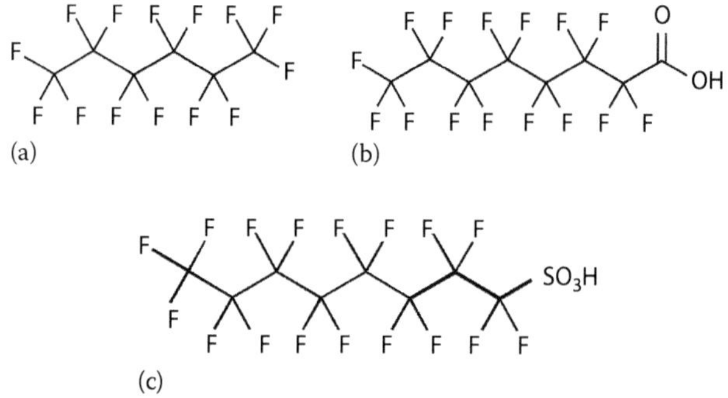

FIGURE 6.3 Perfluorocarbons (a) consist only of carbon (at the interstices) and fluorine atoms. Perfluoroalkyl substances include (b) perfluorooctanoic acid (PFOA) and (c) perfluorooctane sulfonate (PFOS) (c).

(Figure 6.3a). They vary only in the number of carbon atoms linked together. PFCs are used in a variety of industrial applications. They are nontoxic but are a significant contributor to greenhouse gases, which are associated with global climate change. They are extremely stable in the environment. The C–F bond is arguably the strongest single bond encountered in organic chemistry, and its strength is further increased when several fluorine atoms are present on the same carbon atom. The presence of fluorine even reinforces the C–C bonds.

Perfluoroalkyl substances (PFASs)—This is a group of fluorinated chemicals that are entirely man-made. The most common and widely studied chemicals in this group include perfluorooctanoic acid (PFOA), perfluorooctane sulfonate (PFOS), and other perfluoroalkyl substances (PFASs; Figure 6.3b and c). They are all very persistent in both the environment and in organisms. PFASs have been widely used to make products more stain resistant, waterproof, and/or nonstick. In contrast to PFCs, PFASs have toxic effects.

We reserve another family of halogenated organic compounds, organochlorine pesticides, for the next chapter.

6.2 POLYCHLORINATED BIPHENYLS

The general molecular formula for PCBs is $C_{12}H_{10-x}Cl_x$, where x is the number of chlorines (Figure 6.1a). Each of the hydrogen atoms in the biphenyl can be substituted by a chlorine atom. Because the number of sites where chlorine can attach is limited, there is a maximum of 209 different individual PCBs or **congeners**. PCBs are entirely man-made, and not all of the possible congeners have been synthesized in any appreciable amount. Up until the 1970s, PCBs were used for several industrial purposes, mostly associated with cooling transformers and similar devices. Polychlorinated biphenyls are lipophilic and can bioaccumulate, bioconcentrate, and biomagnify in aquatic food chains. The lipophilicity and persistence are closely related to the number of chlorines in the molecule. Some PCBs cause cancer, endocrine disruption, neurotoxicity, and immunotoxicity in animals and are proven carcinogens in humans.

A small group of PCBs are termed **dioxin-like** because their structure causes them to behave like the more toxic dioxins in producing harmful physiological effects (see below). Because of their environmental persistence and suspected human carcinogenicity, PCB production in the United States and several other countries was banned in 1972 and usage was halted in 1979. PCBs have since been included among the POPs recommended to be banned throughout most of the world by the Stockholm Convention in 2001.

6.2.1 CHEMISTRY

On each ring of a biphenyl, there are potentially five sequentially numbered reactive sites (site 1 is occupied by the bond connecting the rings in PCBs, Figure 6.1). If no chlorination occurs, the molecule is simply called **biphenyl**. In PCBs, up to 10 chlorine atoms can be 'substituted' onto the biphenyl, one at each reactive site. The specific congener is given a chemical name based on the number and location of its chlorine atoms. For example, 2,3,3′,5′-tetrachlorobiphenyl is a PCB with four chlorines (hence the prefix *tetra-*) located at sites 2 and 3 on the one phenyl ring and at sites 3 and 5 on the other ring. Each congener is placed in a class based on the number of chlorine ions and given a unique number, so 2,3,3′,5′-tertrachlorobiphenyl is also called PCB 58. PCB 209 is the only PCB that is fully substituted by chlorine atoms.

PCBs can be either **coplanar** or **nonplanar**, and their configuration can affect their toxicity. Imagine two benzene rings in space. If the surfaces of the two rings are in the same plane, that is, a 3D representation would have both lie on a sheet of paper (Figure 6.4a), they are coplanar (sometimes simply called planar). Substitutions at the 2 or 6 positions on either phenyl group are called *ortho*-substitutions, and these can produce torsion that rotates the phenyl rings out of being coplanar to nonplanar (Figure 6.4b). PCBs that do not have any chlorine atoms attached at sites 2 or 6 are referred to as **non-*ortho*-substituted** PCBs. If only one chlorine is attached at either of these sites, the PCB is described as being **mono-*ortho*** substituted.

The level of chlorination in PCBs has substantial effects on their chemical nature. PCBs with three or fewer chlorine atoms are odorless, tasteless, have moderate viscosity, and a clear to pale yellow color. Those with four or more chlorines are deep yellow and viscous. There is a trend for melting points to increase with the degree of chlorination so that at room temperatures most PCBs tend to be solids but some mono- and dichlorophenyls are liquid (Table 6.1). Chlorination decreases water solubility, as seen in both the log K_{ow} values and actual solubilities. PCBs with four or more chlorine atoms have very low solubility in water. Higher chlorination also increases persistence.

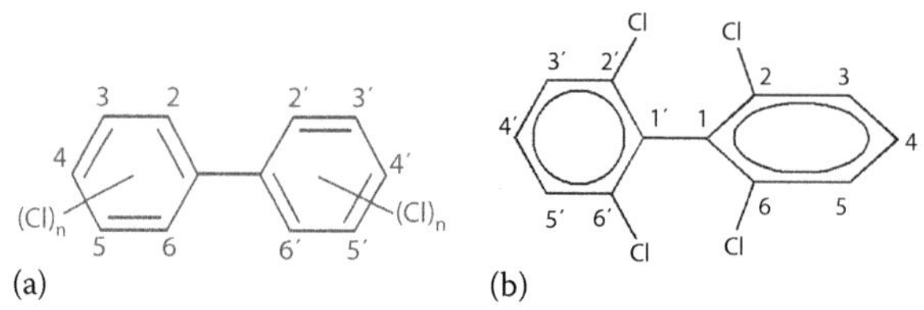

FIGURE 6.4 Examples of coplanar (a) and nonplanar (b) PCBs.

TABLE 6.1

Typical Characteristics of PCBs Based on Number of Chlorine Atoms

Category and No. of Chlorines	Molecular Weight (g)	Solubility (µg/L)	Melting Point (°C)	Half-Life (Years) Soil	Water	Air
Mono (1)	188	5.5	35	0.6	0.17	0.01
Di (2)	225	1.25	25	0.8	0.25	0.02
Tri (3)	257	0.4	45	1.45	0.34	0.06
Tetra (4)	292	0.04	125	5.71	4.01	0.29
Penta (5)	326	0.004	115	4.17	3.09	0.22
Hexa (6)	360	0.0004	101	8.56	9.26	0.66
Hepta (7)	395	0.0004	150	8.56	6.31	0.44
Octa (8)	429	0.0002	160	18.3	3.83	2.97
Nona (9)	464	0.0001	207	20.5	41.5	3.25
Deca (10)	498	1×10^{-6}	226	22.8	68.5	5.7

In 1881 German chemists produced the first PCBs (ATSDR, 2000). In 1929 the Monsanto Chemical Company purchased the rights to produce PCBs and commercial production began in earnest. For more than 40 years, PCBs were widely used as compounds that were effective in keeping electrical transformers cool, as additives in plastics, and in reducing fire hazards. Monsanto blended PCB congeners into 12 commercially sold **Aroclors** that were distinguished by the degree of chlorination. From 1930 to 1977, an estimated 272,700 tonnes of Aroclors were produced in the United States, with an additional 4,204,500 tonnes in Europe (ATSDR, 2000). Of the 209 possible polychlorinated biphenyls, only 130 were manufactured commercially (UNEP, 1999).

6.2.2   PERSISTENCE

Table 6.1 lists some estimated half-lives of PCBs in air, water, and sediment or soil. These are estimates only. Various factors including environmental temperature, acidity, exposure to ultraviolet radiation (sunlight in the environment), and concentration of organic carbons can affect persistence. PCBs in air degrade most rapidly with those having only 1–3 chlorine atoms having half-lives measured in a few days. In contrast, the half-life of PCB 209 in air is 5–6 years. PCBs are more stable in water than in air and usually more stable in soils and sediments than in water.

Water and soil or sediments serve as the primary reservoirs of PCBs (Rice et al., 2003). Although PCBs are not very soluble in water, the vast amounts of salt and freshwater on the planet allow for substantial storage. Even though vapor pressure of PCBs tends to be low, volatilization from sediments and soils is the main way PCBs enter the atmosphere.

6.2.3   BREAKDOWN OF PCBS

Despite their persistence, PCBs will eventually decompose or dechlorinate. They are slowly metabolized by many living organisms (Tehrani and Van Aken, 2014). They can be naturally hydroxylated by organisms through the P450 system, but as we have seen with other compounds, the P450 system can make PCBs more toxic than their parent forms, and this step has the added disadvantage of preventing further decomposition of the molecule. OH-PCBs formed by metabolic activity in living organisms can be released into the environment and enter the food chain through excretion, consumption, and natural cycling of vegetation (Tehrani and Van Aken, 2014). Hydroxylation can occur in the atmosphere through chemical binding with free –OH radicals.

The most common method to remediate soils that are heavily contaminated by PCBs is to mechanically move them to hazard landfills. If the landfills are properly designed and maintained, they can harbor PCBs for decades while they slowly decompose. The technology to reduce the chlorination of the PCBs in these landfills is improving and often relies on enhancing microbial decay of the molecules. Much of the PCBs in water will eventually deposit into sediments where they will be buried and removed from contact with organisms. By turning soil and sediment over, helping to aerate it, and mixing it with organic matter, even lowly earthworms can

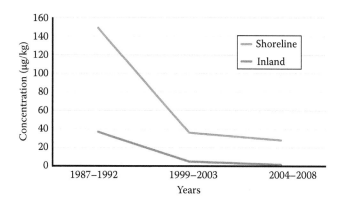

FIGURE 6.5 Data and estimated trend lines for total PCB concentrations (μg/g wet weight) in bald eagle blood plasma, 1987–2006. Eagles living inland had lower PCB concentrations than those living along the shore because of differences in diets. However, both groups of eagles demonstrated significant declines in PCB concentrations over those years. (Data from Wierda, M.R. et al., *Environ. Toxicol. Chem.*, 35, 1995, 2016.)

accelerate the degradation of PCBs (Singer et al., 2001). Certain species of plants can be used to take up PCBs from soils or sediments.

Because PCBs are no longer manufactured or used in this or other countries, their concentrations are diminishing in many environmental sources. For example, Weirda et al. (2016) found that between 1987 and 2008, PCB concentrations dropped by 80%–81% in bald eagles (*Haliaeetus leucocephalus*) living both along the coast of the Great Lakes and inland (Figure 6.5). Similarly, Buehler et al. (2002) reported that atmospheric PCBs declined by 91% between 1978 and 2000 along Lake Superior.

6.2.4 CONCENTRATIONS OF PCBs IN SOME ANIMAL TISSUES

PCBs can bioconcentrate and biomagnify due to their lipophilic and persistent natures. For instance, the average congener concentration in oysters in Manila Bay, Philippines, was 20–60 times higher than in sediments (Villeneuve et al., 2010, Figure 6.6). Oysters may obtain some of their contaminant burdens through their food, but they undoubtedly obtain much of it directly from the sediments.

Concentrations of specific congeners vary predictably due to characteristics of the congener, species of concern, and location. Very low molecular weight congeners will volatize, which may make them unavailable to most organisms. The heaviest congeners are mostly immobile in the soil or sediments, so they too are less common in organisms. Therefore, middle-weight congeners obtain the highest concentrations in animals and plants and present the greatest toxicological concern. PCBs concentrate in lipids, so animals that have more fat may have higher concentrations of PCBs, all other factors being the same. Because of the high affinity between lipids and PCBs, studies often report concentrations in lipid-corrected values. That is, total PCBs (perhaps by congener) will be measured along with percent body fat, and the concentration will be shown as concentration per unit of lipid.

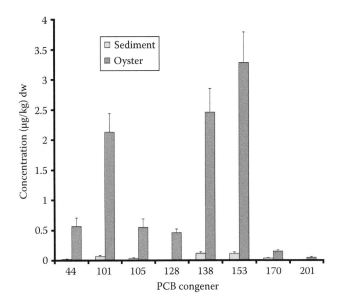

FIGURE 6.6   An example of bioconcentration of PCBs between oysters and sediments in the Philippines. The composition of congeners is similar between sediments and oysters. (Adapted from Villeneuve, J.P. et al., *Int. J. Environ. Health Res.*, 20(4), 259, 2010.)

Marine mammals often have comparatively high concentrations of PCBs, in part because of the large amount of fat they need for insulation. Also, the metabolic potential to degrade organochlorine contaminants is lower in cetaceans than in many terrestrial animals (Tanabe et al., 1987). For example, total concentrations in the striped dolphin (*Stenella coeruleoalba*) from the Mediterranean Sea in the 1990s were as high as 855 mg/kg, among the highest reported in the literature (Eisler, 2000). Kannan et al. (1993) determined that about 53% of the total PCBs in striped dolphins was composed of dioxin-like coplanar congeners having substantially higher toxicities than other PCBs. Studies on other cetaceans and pinnipeds usually report total PCB concentrations from the low µg/kg range up to 50 mg/kg lipid weight.

Some fatty fish have high concentrations of PCBs, depending on location. Marine-dwelling flounder collected from contaminated waters had 124–333 mg/kg total PCBs dry weight, whereas those from a reference area had only 4–13.3 mg/kg. Lake trout (*Salvelinus namaycush*), a top freshwater predator, had up to 8 mg/kg lipid weight total PCBs in 1977, but this gradually declined to 2.5 mg/kg by 1988 (Borgmann and Whittle, 1991). Within a species, significant differences among sexes or age classes can occur due to diet, lipid content, and age. Those animals that eat a lot of fish, have higher body fat (as is often the case with female mammals), or are older tend to have higher concentrations than the alternatives. Fish advisories warning people to limit or cease consuming fish from particular lakes due to high levels of contaminants including PCBs in their tissues exist for several states in the Eastern and Midwestern United States.

Maternal transfer is a concern for animals and humans alike. This can occur when PCBs in the mother are transported into embryos or nursing young. For egg-laying animals, egg yolks have high concentrations of fats, which can convey lipid-soluble PCBs and other organic contaminants. PCBs can also be transferred in mammalian milk. Fortunately, concentrations in human milk seem to be declining in parts of the world. For example, Ryan and Rawn (2014) demonstrated that PCBs in human milk in Canada have declined by 11% between 2002 and 2010.

6.2.5 BIOLOGICAL EFFECTS OF PCBs

Generally, PCBs exert very low toxicity to plants because the more toxic, highly chlorinated PCBs, including those having dioxin-like properties, are not assimilated very well by most plants. When toxicity is demonstrated, it is usually evident by slower growth rates or smaller leaves than control plants.

We mentioned that plants are used to remediate contaminated soils. For instance, Viktorova et al. (2014) genetically modified tobacco plants by adding a bacterial gene that produced an enzyme that breaks the biphenyl ring. Genetically modified plants were more tolerant to PCB-contaminated soils than nonmodified plants, and they were more efficient in reducing PCB concentrations in a dumpsite. The common pumpkin (*Cucurbita pepo*) and related squashes and gourds seem to assimilate PCBs at a level that would make them useful in soil remediation (Zeeb et al., 2006).

In the aquatic copepod, *Daphnia magna*, 24 hour LC50s range from 710 to 860 μg/L in a mixture of very low-chlorinated PCBs, but toxicity increased with degree of chlorination (Dillon and Burton, 1992). Lethal concentrations in fish range in the tens of thousands of μg/kg (reviewed by Eisler, 2000). Considerable variation in toxicity exists among congeners and species of birds, making it difficult to arrive at a meaningful range of LD50 or LC50 values. Egg injection studies suggest that some avian embryos can tolerate 400 or more μg/kg egg weight but domestic chickens succumb at 0.6 μg/kg with the more toxic, dioxin-like congeners (reviewed by Eisler, 2000). Median lethal doses (LD50) for rats fed with PCBs ranged from 1010 to 4250 mg/kg in the food; values for mink (*Mustela vison*) were 730 to 4000 mg/kg. Chronic lethality occurred at 7.1 mg/kg/day over 28 days in mink. Sublethal effects were seen at doses of 1 mg/kg food/day in rats, mice, and mink in various organ systems. The liver is a common target for PCB toxicity due to its high rate of P450 activity. PCBs also affect thyroid gland hormone production and histology at low doses. Toxic effects among Aroclor mixtures differed significantly in many studies, with higher degrees of chlorination exerting stronger effects than lower chlorination (reviewed by ATSDR, 2000).

6.3 ECOTOXICITY OF DIOXINS, FURANS, AND DIOXIN-LIKE COMPOUNDS

Both dioxins and furans come from several natural and human-related sources that typically involve high heat conditions. Volcanoes, forest fires, brush fires, the normal processing of sugarcane through burning prior to harvesting (Figure 6.7), and burning of other biomass are major natural sources of these compounds. Industrial combustion processes such as garbage incineration can also produce substantial dioxins

FIGURE 6.7 Burning of sugarcane and other forms of combustion are ways that dioxins and furans are formed due to human activities. (Courtesy of Wikimedia Commons, Forster, P., CC BY-SA 2.0, http://creativecommons.org/licenses/by-sa/2.0.)

and furans. Accidental fires or breakdowns involving capacitors, transformers, and other electrical equipment (e.g., fluorescent light fixtures) that contain PCBs can release furans and dioxins formed by thermal processes.

Dioxins and furans are structurally very similar to PCBs, and as a result, the groups have similar properties of solubility in oils, low water solubility, effects of chlorination on melting points, stability, and generally strong environmental persistence. Dioxins and furans structurally differ from PCBs in that the former link the two phenyl groups with oxygen atoms whereas the later do not; moreover, dioxins have two oxygen atoms between the phenyl groups, whereas furans have only one (Figure 6.1b and c). Due to chlorine substitutions, there are a maximum of 75 dioxins, with 7 of these being highly toxic. Furans have 135 congeners, 10 of which have highly toxic dioxin-like properties.

Some of the more salient characteristics of dioxins and furans are mentioned in Table 6.2. Low molecular weight molecules tend to have higher solubility and lower melting points than heavier molecules. All dioxins have low vapor pressures, which is a predictor of how likely they are to be in the air; higher vapor pressures indicate a greater tendency to be in the atmosphere. Various factors affect half-lives of dioxins and furans, but values generally run in a few hours in the atmosphere, from a few days to half a year in water, and at least 20 years in soils and sediments for dioxins and furans with four or more chlorine atoms. Furans are very similar to the values reported here.

6.3.1 General Mechanisms of Toxicity

Dioxins and dioxin-like congeners of PCBs and furans are highly toxic chemicals that are linked or known to produce cancer, endocrine disruption, neurological disorders, liver toxicity, heart disease, and genotoxicity in animals and presumably in humans (ATSDR, 1998). Animals are usually exposed through the food chain, and being part of that food chain is how most humans also become exposed.

TABLE 6.2

Approximate Values for Some Characteristics of Dioxins Based on Degree of Chlorination

Number of Chlorines	Molecular Weight (g/mol)	Solubility (mg/L)	Melting Point (°C)	Vapor Pressure (Pa)
1	218	0.318	90	0.015
2	253	0.149	164	0.00014
3	287	0.841	129	0.0001
4	322	3.2×10^{-3}	190	6×10^{-6}
5	356	1.2×10^{-3}	196	8.8×10^{-7}
6	391	4×10^{-6}	273	5.1×10^{-9}
7	425	2×10^{-6}	265	7.5×10^{-10}
8	460	7×10^{-7}	332	1.1×10^{-10}

Source: Shiu, W.Y. et al., *Environ. Sci. Technol.*, 22, 651, 1988.

So what makes a PCB or furan dioxin-like? Of the 209 possible PCB congeners, only 12 are highly toxic due to their dioxin-like structure; similarly, only a few furans and even a selection of dioxins are highly toxic. This is due to the commonality in their structure. The common characteristic of the highly toxic dioxins, furans, and PCBs with dioxin-like qualities is that they can bind with the **aryl hydrocarbon receptor** in animals and through this exert their toxic effects. The receptor is internal to cells and is activated by attachment to certain organic molecules. Once activated, it facilitates the P450 system, so it has broad-ranging effects. From a structural perspective, molecules that are chlorinated at the 2 and 3 or 7 and 8 positions on the phenyl rings attach most readily to the aryl hydrocarbon receptor. The most toxic of dioxins, 2,3,7,8-tetrachlorodibenzo-p-dioxin (TCDD) has both active sites and with only four chlorines is mobile and readily assimilated (Figure 6.8). Because organisms are typically exposed to many dioxin-like molecules simultaneously and the relative toxicity of dioxin-like chemicals varies considerably, the World Health Organization (WHO) developed the concept of **toxic equivalency quotient (TEQ)** and **toxic equivalency factor (TEF)** (Van den Berg et al., 2006). Each dioxin-like compound has a TEF and the mixture of dioxin-like compounds is assigned a TEQ based on the sum of TEFs.

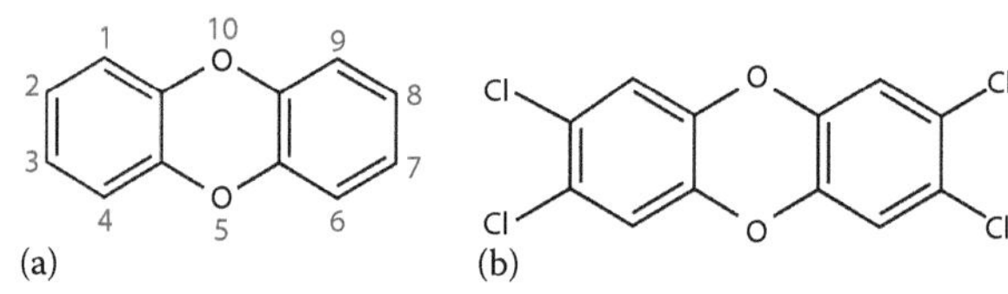

FIGURE 6.8   An illustration showing dibenzodioxin (a) with the substitution sites numbers and 2,3,7,8-tetrachlorodibenzo-p-dioxin (TCDD), the most toxic of the dioxins. (b) Chlorine atoms attached to sites 2 and 3 or 7 and 8 increase the ability of the dioxin-like compound to attach to a receptor of cells and cause numerous physiological problems.

TCDD is given a TEF of 1 and the toxicity of all other chemicals is a factor of this. Thus, PCB 126 has a TEF of 0.1, suggesting that its toxicity is a tenth of that of TCDD. Dioxins other than TCDD, furans, and non-ortho substituted PCBs have TEFs ranging from 0.1 to 0.0003. Nondioxin PCBs have much lower toxicity, about 0.00003 as toxic as TCDD. This system was developed as an estimate for human toxicity and may not equally apply to all species, although it has been used in many different scenarios. It is safe to conclude that a TEQ of 2 is going to be more toxic than a TEQ of 0.2, but to suggest that the first is exactly 10 times more toxic than the second is stretching the limits of the system. It is also true that the toxicity of most dioxin-like PCBs is low to very low compared to TCDD.

Many toxicological studies have examined the effects of TCDD on a variety of organisms. A single 3–5 µg/kg body weight of this congener led to the LC50 in zebrafish (*Danio rerio*) and 5.6 µg/kg exposure for 96 hour was the LC50 in coho salmon (*Oncorhynchus kisutch*). A diet of 1.7 ng/kg body weight resulted in decreased reproduction, abnormal ovarian histology, and developmental effects in sensitive species (reviewed by Eisler, 2000). For birds, dietary LD50 values ranged from 15 µg/kg body weight in northern bobwhite (*Colinus virginianus*) to 108 µg/kg in mallards (*Anas platyrhynchos*) and 810 µg/kg body weight in ringed-collared doves (*Streptopelia risoria*). Among mammals, LD50 values ranged from 2.5 µg/kg body weight in guinea pigs (*Cavia* sp.) to 1157–5051 µg/kg body weight in hamsters (*Cricetus* sp.). In perspective, LD50s in guinea pigs ranged from 2 µg/kg body weight with 2,3,7,8-TCDD to more than 300,000 µg/kg body weight with 2,8-DCDD (Kociba and Schwetz, 1982).

6.4 POLYBROMINATED DIPHENYL ETHERS AND POLYBROMINATED BIPHENYLS

Instead of chlorine, bromine substitutes for hydrogen on the phenyl rings of these chemicals. PBDEs and PBBs (Figure 6.2a and b) were or are used as flame retardants in a wide variety of industries including textiles, automotives, building materials, electronics, furnishings, aeronautics, and plastics. They share similar characteristics of persistence, bioconcentration, biomagnification, and lipophilicity and effects of increasing halogenation with the other chemicals mentioned in this chapter. Because they have the same ring structure as PCBs, both PBDEs and PBBs have 209 congeners. Commercially marketed PBDEs were sold as mixtures of several congeners.

As intimated in the opening paragraphs of this chapter, PBBs were first voluntarily removed from the market by their manufacturer in the 1970s due to a major accident involving thousands of livestock. In early 1973, both PBB (sold under the trade name FireMaster®) and magnesium oxide (a cattle feed supplement sold under the trade name NutriMaster®) were produced at the same St. Louis, MI, plant by the Michigan Chemical Company. About 20 fifty pound bags of PBB were accidentally sent to Michigan Farm Bureau Services in place of NutriMaster. The error was not discovered until after the bags had been shipped to feed mills and used in the production of livestock foods, nearly a year later. By the time the mistake became apparent, PBB had entered the food chain through milk and other dairy products, beef products, and contaminated swine, sheep, chickens, and eggs. As a result, over 500 contaminated Michigan farms were quarantined and approximately 30,000 cattle, 4,500 swine, 1,500 sheep, and 1.5 million chickens were destroyed, along with over 800 tons of animal feed, 18,000 lb of cheese, 2,500 lb of butter, 5 million eggs, and 34,000 lb of dried milk products (Michigan Department of Public Health, 2011). While this may seem as 'ancient' history, it is not beyond imagination that a similar accident could occur again with similar compounds. The manufacture of PBB was banned in the United States in 1976 after this catastrophe (EPA, 2014).

PBBs affect just about every organ system in animals, but thyroids and livers are principal targets. LD50 values for rats included a 90-day study with 65–149 mg/kg/day in rats and a 313-day study in mink with 0.47–0.61 mg/kg/day. As with PCBs, environmentally relevant exposures with PBBs must occur over a prolonged period (ATSDR, 2004).

PBDEs are currently used primarily as flame retardants. They are persistent although the bromine–carbon bond is weaker than the chlorine bond and PBDEs are not quite as persistent as their chlorinated counterparts. They can bioaccumulate and have been linked to cancer, thyroid problems, and neurological disorders (ATSDR, 2015). Since the 1970s, PBDEs have been used in electronics, textiles, and polyurethane foam because they can elevate the combustion temperatures of these products and make them more fire resistant.

Manufacturers produced penta-, octa-, and deca-congeners of the family, but deca accounts for about 80% of the total production. All three congeners were voluntarily taken off of the market in the European Union, the octa- and penta-congeners have been banned by the EPA in the United States in 2004, and deca production is being voluntarily phased out, starting since 2010 (EPA, 2014). Due to stockpiles of the chemical, decabrominated biphenyl ether is still used, but some states have banned or greatly restricted its use.

Serious health effects have been associated with exposure to PBDEs, mostly in laboratory and especially with low or moderate levels of bromination. However, deca-BDEs should not cause serious effects at environmentally relevant concentrations (ATSDR, 2015). An LD50 of 5000 mg/kg was measured in rats with a mixture of penta-BDEs, but no deaths occurred at either octa- or deca-BDEs at the same dose. Moreover, no deaths were observed with deca-BDEs at concentrations less than 8 mg/kg during a 13-week exposure. However, endocrine effects occur at much lower dose levels. Oral doses of 10 mg/kg/day over a month can significantly reduce serum thyroxine concentrations (ATSDR, 2015). PBDEs can biomagnify. Other sublethal effects of PBDEs include neurotoxicity, developmental disorders, cancer, impaired learning and memory, affected sexual development and other behaviors, and negative effects on the liver and thyroid (ATSDR, 2015). PBDEs are virtually everywhere, including the Arctic and the Antarctic and in eggs and fetuses through material transfer. The purpose of PBBs and PBDEs—to provide protection against fire—is still of concern and other brominated and organophosphate compounds are being used to fill that need.

6.5 POLYFLUORINATED ORGANIC COMPOUNDS

We will dedicate less space to these molecules than those above for they do not have, with one exception, the history of causing environmental problems as the others do, are not as widely used, and have no to limited toxicity. While fluorine atoms can be combined with biphenyls, they are only used in specialty commercial applications and at a much lower production volume than necessary to cause much of a problem.

That leaves two classes of fluorinated organic compounds that we need to be concerned about at all—those that consist only of carbon and fluorine, perfluorocarbons, and those that have carbon, fluorine, and other groups such as hydroxyls or sulfides, the so-called perfluoroalkyl substances.

Perfluorocarbons are only distinguished from each other by their length. These chemicals have shorthand notation of C_xF_y. They are extremely stable, for the carbon–fluorine bond is one of the strongest in organic chemistry due to a few highly technical reasons that involve mutual strengthening of the bonds. Perfluorocarbons are more chemically and thermally stable than their corresponding hydrocarbon counterparts, and indeed any other organic compound (O'Hagan, 2008). They can succumb to very strong reducing agents and some organometallic complexes.

Perfluorocarbons are colorless. They are both lipophobic and hydrophobic, meaning that they do not mix well with lipids, organic solvents, or water. Due to the strong internal bonds, perfluorocarbons are essentially inert. They can be very volatile, especially the lighter molecular weight molecules. The smaller ones are also liquid at room temperature.

Perfluorocarbons are used in polymers and as refrigerants, solvents, and anesthetics and for other medical purposes. When synthesized with chlorine atoms, the resulting chlorofluorocarbons or CFCs have caused problems as ozone-depleting refrigerants. CFCs were little used for refrigeration until better synthesis methods, developed in the 1950s, reduced their cost. Their domination of the market was called into question in the 1980s by concerns about depletion of the ozone layer. Legislation on greenhouse gases banned the use of most CFCs, and hydrochlorofluorocarbons (HCFCs) and perfluorocarbons were used as substitute refrigerants. Now, however, these chemicals are being scrutinized under the Kyoto Protocol on climate change for the possible ozone-depleting characteristics. Because perfluorocarbons are inert and essentially cannot be metabolized, they have no known toxic properties.

As mentioned, perfluoroalkyl substances (PFASs) are entirely man-made. The most common and widely studied chemicals in this group include perfluorooctanoic acid (PFOA), perfluorooctane sulfonate (PFOS), and other PFASs (Figure 6.3b and c). They are all very persistent in both the environment and in organisms. PFASs have been widely used to make products more stain resistant, waterproof, and/or nonstick. Teflon® is a PFOS, as is Scotchgard®. Other uses of PFOSs are as agents for textiles, paper, and leather and in wax, polishes, paints, varnishes, and cleaning products for general use. In another industry, PFOS is used in photolithographic chemicals, including antireflective coatings. Due to its extreme stability in the environment, PFOS has been declared a POP under the Stockholm Convention. It has been found throughout the world and in animals as diverse as albatrosses and polar bears (*Ursus maritimus*).

Perfluorooctane is toxic to animals. It has a 96 hour LC50 of 251 mg/kg in rats (EPA, 2016). Sublethal effects include decreased body and liver weights, liver damage, and decreased cholesterol at concentrations as low as 0.5 mg/kg. There is some indication of carcinogenic tumor production at concentrations low enough to reflect human exposure.

Perfluorooctanoic acid is similarly very stable and long-lived in the environment. It is toxic and carcinogenic in both animals and humans. Over 98% of the general U.S. population has traces of PFOA in their blood at sub µg/kg concentrations. As might be expected, concentrations are higher in chemical plant employees and in populations around plants. PFOA has been detected in industrial waste, stain-resistant carpets, carpet cleaning liquids, house dust, microwave popcorn bags, water, food, some cookware, and PTFE such as Teflon.

Companies that manufacture either PFOS or PFOA are entering agreements with the EPA to voluntarily cease production of the chemicals, and as a result, these chemicals are on their way out.

6.6   CHAPTER SUMMARY

1. Most of the contaminants highlighted in this chapter have a common structural feature of two phenyl groups attached to each other. Polychlorinated and polybrominated biphenyls have a straight connection between adjoining carbon atoms; polybrominated diphenyl ethers, dioxins, and furans have the two phenyl groups connected through one or two oxygen atoms. The fluorinated compounds may exist as cyclic compounds or in chains of carbon atoms.
2. All the molecules are very persistent and soluble in lipids, and most can both bioconcentrate and biomagnify.
3. These contaminants can cause several sublethal but serious effects, including cancer, deformities, endocrine disruption, and liver and thyroid diseases.
4. Toxicity varies considerably among the congeners in biphenyl-based chemicals. Certain chlorinated contaminants are called dioxin-like because of their configuration that allows them to bind with the aryl hydrocarbon receptor.
5. PCBs, PBBs, and PBDEs are either being phased out in different parts of the globe or have been banned altogether. Dioxins and furans occur naturally due to high-temperature combustion processes, and they can be created during industrial operations, but they have no known practical application and are not synthesized in large quantities.
6. Fluorinated organic compounds are extremely persistent due to the carbon–fluorine bonds. They range from being completely nontoxic to causing cancers and liver diseases.

6.7   SELF-TEST

1. Which of the following compounds are or were manufactured for industrial purposes?
   a. Dioxins
   b. Polychlorinated biphenyls
   c. Furans
   d. More than one above

2. Structurally, which two molecular families are most similar?
   a. PBDEs and dioxins
   b. PCBs and PBBs
   c. Furans and PCBs
   d. PBDEs and PCBs
3. What meeting identified certain contaminants as persistent organic pollutants?
   a. Helsinki Accord
   b. Stockholm Convention
   c. Stockholm Syndrome
   d. United Nations Environmental Programme (UNEP)
4. Which of the following structural formulas is for a polybrominated biphenyl?
   a.

   b.

   c.

   d.

5. As the molecular weight of PCBs increases, which of the following characteristics changes?
   a. Melting points increase
   b. Solubility decreases
   c. Half-lives increase
   d. More than one above
6. Major natural sources of dioxins and furans include
   a. Volcanoes
   b. Freshwater lakes
   c. Icebergs
   d. There are no natural sources of dioxins or furans
7. A solution of dioxin-like contaminants has individual chemicals with toxic equivalency factors (TEF) of 0.002, 0.01, 0.1, and 0.0003. What is the toxic equivalency quotient (TEQ) of the solution?
   a. 0.1123
   b. 0.1
   c. 0.003
   d. The answer cannot be solved from the data provided.

8. Polybrominated diphenyl ethers are still being used in the United States as what application?
   a. Additives to milk shakes
   b. Production of steel
   c. Flame protectants
   d. In the making of pesticides
9. Short answer. Explain why PCBs and PBBs have exactly 209 possible congeners.
10. True or False. All of the fluorinated organic compounds mentioned here are completely safe and nontoxic.

REFERENCES

Agency for Toxic Substances and Disease Registry (ATSDR) 1998. *Toxicological Profile for Chlorinated Dibenzo-p-Dioxins (CDDs)*. Atlanta, GA: U.S. Department of Health and Human Services.

Agency for Toxic Substances and Disease Registry (ATSDR) 2000. *Toxicological Profile for Polychlorinated Biphenyls (PCBs)*. Atlanta, GA: U.S. Department of Health and Human Services.

Agency for Toxic Substances and Disease Registry (ATSDR) 2004. *Toxicological Profile for Polybrominated Biphenyls (PCBs)*. Atlanta, GA: U.S. Department of Health and Human Services.

Agency for Toxic Substances and Disease Registry (ATSDR) 2015. *Draft Toxicological Profile for Polybrominated Brominated Diphenyl Ethers (PBDEs)*. Atlanta, GA: U.S. Department of Health and Human Services.

Borgmann, U., Whittle, D.M. 1991. Contaminant concentration trends in Lake Ontario lake trout (*Salvelinus namaycush*)—1977 to 1988. *J Great Lakes Res* 17: 368–381.

Buehler, S.S., Basu, I., Hites, R.A. 2002. Gas-phase polychlorinated biphenyl and hexachloro-cyclohexane concentrations near the Great Lakes: A historical perspective. *Environ Sci Technol* 36: 5051–5056.

Dillon, T.M., Burton, W.D.S. 1992. Acute toxicity of PCB congeners to *Daphnia magna* and *Pimephales promelas*. *Bull Environ Contam Toxicol* 46: 208–215.

Eisler, R. 2000. Polychlorinated biphenyls. Pp. 1237–1341. In: Eisler, R. (ed.), *Handbook of Chemical Risk Assessment: Health Hazards to Humans, Plants and Animals*, Vol. 2, Organics. Boca Raton, FL: Lewis Publishers.

Kannan, N., Tanabe, S., Borrell, A., Aguilar, A., Focardi, S., Tatsukawa, R. 1993. Isomer-specific analysis and toxic evaluation of polychlorinated biphenyls in striped dolphins affected by an epizootic in the western Mediterranean Sea. *Arch Environ Contam Toxicol* 25: 227–233.

Kociba, R.J., Schwetz, B.A. 1982. A review of the toxicity of 2,3,7,8-tetrachlordibenzo-p-dioxin (TCDD) with a comparison to the toxicity of other chlorinated dioxin isomers. *Assoc Food Drug Off Quart Bull* 46: 168–188.

Michigan Department of Public Health. 2011. PBBs (polybrominated biphenyls) in Michigan: Frequently asked questions—2011 update. https://www.michigan.gov/documents/mdch_PBB_FAQ_92051_7.pdf. Accessed July 9, 2016.

O'Hagan, D. 2008. Understanding organofluorine chemistry. An introduction to the C–F bond. *Chem Soc Rev* 37: 308–319.

Rasmussen, P.W., Schrank, C., Williams, M.C.W. 2014. Trends of PCB concentrations in Lake Michigan coho and chinook salmon, 1975–2010. *J Great Lakes Res* 40: 748–754.

Rice, C.P., O'Keefe, P.W., Kubiak, T.J. 2003. Sources, pathways, and effects of PCBs, dioxins, and dibenzofurans. Pp. 501–573. In: Hoffman, D.J., Rattner, B.A., Burton Jr., G.A., Cairns Jr., J. (eds.), *Handbook of Ecotoxicology*. Boca Raton, FL: Lewis Publishers.

Ryan, J.J., Rawn, D.F.K. 2014. The brominated flame retardants, PBDEs and HBCD, in Canadian human milk samples collected from 1992 to 2005; concentrations and trends. *Environ Int* 70: 1–8.

Shiu, W.Y., Doucette, W., Gobas, F.A.P.C., Andren, A., Mackay, D. 1988. Physical-chemical properties of chlorinated dibenzo-p-dioxins. *Environ Sci Technol* 22: 651–658.

Singer, A.C., Jung, W., Luepromchai, E., Yahng, C.-S., Crowley, D.E. 2001. Contribution of earthworms to PCB remediation. *Soil Biol Biochem* 33: 765–776.

Tanabe, S., Kannan, N., Subramanian, A., Watanabe, S., Tatsukawa, T. 1987. Highly toxic coplanar PCBs: Occurrence, source, persistency and toxic implications to wildlife and humans. *Environ Poll* 47: 147–163.

Tehrani, R., Van Aken, B. 2014. Hydroxylated polychlorinated biphenyls in the environment: Sources, fate and toxicities. *Environ Sci Poll Res* 21: 6334–6345.

United Nations Environmental Programme (UNEP) 1999. *Guidelines for the Identification of PCBs and Materials Containing PCBs*. Geneva, Switzerland. https://wedocs.unep.org/rest/bitstreams/13438/retrieve. Accessed December 9, 2016: UNEP Chemicals.

U.S. Environmental Protection Agency (EPA). 2014. Technical fact sheet—Polybrominated diphenyl ethers (PBDEs) and polybrominated biphenyls (PBBs). https://www.epa.gov/sites/production/files/2014-03/documents/ffrrofactsheet_contaminant_perchlorate_january2014_final_0.pdf. Accessed July 8, 2016.

U.S. Environmental Protection Agency (EPA). 2016. Health effects support document for perfluorooctane sulfonate (PFOS). Washington, DC: U.S. EPA Office of Water. https://www.epa.gov/sites/production/files/2016-05/documents/hesd_pfos_final-plain.pdf. Accessed December 11, 2016.

Van den Berg, M., Birnbaum, L.S., Denison, M., De Vito, M., Farland, W. et al. 2006. The 2005 World Health Organization reevaluation of human and mammalian toxic equivalency factors for dioxins and dioxin-like compounds. *Toxicol Sci* 93: 223–241.

Villeneuve, J.P., Cattini, C., Bajet, C.M., Navarro-Calingacion, M.F., Carvalho, F.P. 2010. PCBs in sediments and oysters of Manila Bay, the Philippines. *Int J Environ Health Res* 20(4): 259–269.

Viktorova, J., Novakova, M., Trbolova, L., Vrchotova, B., Lowecka, P., Makova, M., Macek, T. 2014. Characterization of transgenic tobacco plants containing bacterial bphC gene and study of their phytoremediation ability. *Int J Phytoremediation* 16: 937–946.

Wierda, M.R., Leith, K.F., Roe, A.S., Grubb, T.J. 2016. Using bald eagles to track spatial (1999–2008) and temporal (1987–1992, 1999–2003, and 2004–2008) trends of contaminants in Michigan's aquatic ecosystems. *Environ Toxicol Chem* 35: 1995–2002.

Zeeb, B.A., Amphlett, J.S., Rutter, A., Reimer, K.J. 2006. Potential for phytoremediation of polychlorinated biphenyl-(PCB-)contaminated soil. *Int J Phytoremediation* 8: 199–221.

7 Other Major Organic Contaminants

7.1 INTRODUCTION

In this chapter we cover two more families of organic contaminants. One of these principally occurs naturally but can be man-made; the other is only synthesized. Both can be present in the environment. Those in the first family can be very short-lived and are continually being produced, whereas those in the second family are long-lived and still can be found because of this persistence. Both families are poly-cyclic, meaning that they have a backbone formed by joined phenol rings; the first group is totally hydrocarbon, while the second contains chlorine. Both families have members that are extremely toxic and others that are essentially nontoxic at meaningful concentrations. The two families are polycyclic aromatic hydrocarbons (PAHs) and organochlorine pesticides (OCPs), respectively. We place them here to complete the list of contaminants of greatest concern. During our discussion of PAHs, we will also include oil pollution because these spills are important sources of PAHs.

7.2 POLYCYCLIC AROMATIC HYDROCARBONS

Polycyclic aromatic hydrocarbons, also known as **polyaromatic hydrocarbons**, are a group of more than a hundred different chemicals characterized by two or more aromatic or phenol rings connected to each other and having no other atoms besides carbon and hydrogen (Figure 7.1). Some of these molecules have been joined with other atoms or molecules such as nitrogen, oxygen, hydroxyls (–OH), and sulfhy-dryls (–SH). PAHs may be man-made or naturally occurring. Natural PAHs come primarily from the combustion of organic materials such as forest, brush, or grassland fires. After that come volcanoes, other sources of combustion, and aerobic bacteria. Human-derived sources of PAHs include industrial combustion; burning of fossil fuels; petroleum-related mishaps including oil spills, vehicular exhausts, asphalts, tars, and refuse incineration; and other industrial processes.

Some PAHs are among the most carcinogenic contaminants in the environment (ATSDR, 2009). Single, microgram levels of exposure to certain PAHs can produce carcinogenic tumors in animals. All vertebrates are prone to developing cancer due to PAH exposure, and several PAHs have been declared probable human carcinogens by the Environmental Protection Agency (EPA) and other agencies. These molecules produce several other toxic effects (ATSDR, 2009).

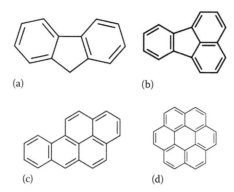

FIGURE 7.1   Examples of polycyclic aromatic hydrocarbons (PAHs): (a) fluorene; (b) fluoranthene; (c) benzo[a]pyrene; and (d) coronene.

Sixteen PAHs are listed by the EPA as **priority pollutants** in drinking water (EPA, 2016a) based on widespread distribution, potential environmental concentrations, and toxicity. These include the following:

Acenaphthene	Acenaphthylene
Fluoranthene	Anthracene
Naphthalene	Benzo[g,h,i]perylene
Benzo[a]anthracene	Fluorene
Benzo[a]pyrene	Phenanthrene
Benzo[b]fluoranthene	Dibenzo[a,h]anthracene
Benzo[k]fluoranthene	Indeno[1,2,3-cd]pyrene
Chrysene	Pyrene

The Agency for Toxic Substances and Disease Registry (ATSDR, 2016) adds benzo[e]pyrene, benzo[j]fluorine, and coronene on its priority list of hazardous substances.

7.2.1   Chemical Characteristics of PAHs

As mentioned above, the most ecotoxicological interest in PAHs is on molecules that are saturated with hydrogen atoms (Figure 7.1). Substituted PAHs, those with other atoms in addition to hydrogens, are less common in the environment but may be more important in industrial settings. Unless otherwise stated, our discussions will focus on saturated PAHS. Environmentally important PAHs range from light molecular weight (128 g) two-ring naphthalene to heavier (300 g) seven-ring coronene.

Like polychlorinated biphenyls (PCBs) and polychlorinated dibenzofurans (PCDFs) (see Chapter 6), as molecular weight and the number of rings increase, PAH molecules become less soluble in water and less volatile, have higher melting points, and are more stable (Table 7.1). Lightweight PAHs such as naphthalene volatilize in air and are relatively water soluble compared to many of the organic molecules

TABLE 7.1

Some Characteristics of Selected Polycyclic Aromatic Hydrocarbons

PAH	Rings	MW	Melting Point (°C)	Solubility in Water (mg/L, 25°C)	Vapor Pressure (mm, 25°C)	K_{ow}
Naphthalene	2	128.2	78	31.8	0.05	3.3
Acenaphthylene	3	154.2	93	4	NA	3.7
Anthracene	3	178.2	216	0.04	1.9E−4	4.6
Chrysene	4	228.3	254	1.0E−3	6.2E−9	5.9
Benzo[a]pyrene	5	252.3	179	2.0E−4	5.5E−9	6.0
Benzo[g,h,i]perylene	6	276.4	278	3.0E−4	1E−10	6.8
Coronene	7	300.3	438	1.4E−4	1.5E−11	7.56

we have examined. If you have ever smelled commercial moth balls and know how pungent they are, you are smelling naphthalene. When we get to four or more rings, however, water solubility is practically nonexistent and the PAH has very low volatility.

7.2.2 Sources and Uses of PAHs

Globally, approximately 520,000 tonnes of PAHs are emitted into the atmosphere each year (Zhang and Tao, 2009). Of that, combustion of biofuels accounts for about 57% and wildfires 17%. The principal contributor to this load is China, followed by India and the United States. An additional 209,000 tonnes are released into aquatic environments each year (Eisler, 2000). While technological advances have improved vehicle and incinerator emissions, the extensive brush and forest fires in the western United States over the past few years have generated additional PAH loads.

The production of PAHs occurs during incomplete combustion of organic materials and typically requires temperatures exceeding 700°C. These high temperatures are easily attained in volcanoes, forest fires, and industrial processes. However, some PAHs can accumulate over a long period of time with combustion at lower temperatures as witnessed by the concentration of PAHs in fossil fuels (Eisler, 2000) and by the natural production of PAHs through bacterial action. Few PAHs mix with water at all; rather, they reside as surface films or layers that eventually cling to particulates and may sink to sediments where they persist for many years.

Due to their high affinity for particulates such as organic matter, land-based PAHs are quickly absorbed by soil where they are subject to bacterial decomposition. In both sediments and soil, sorption greatly reduces the availability of PAHs to organisms and thus reduces their risk of toxicity.

7.2.3 Persistence

In comparison to other organic pollutants, loss of PAH from the environment can be rather rapid. Factors that influence the rate of decay include oxygen, pH, and particle size of the soil or sediment. Oxygen supports aerobic bacteria and makes decay more efficient. Breakdown of PAHs occurs most rapidly at neutral or slightly alkaline soil or water. Small particle size slows PAH loss compared to larger size such as sand.

Volatile PAHs are particularly sensitive to ultraviolet light and may last only a few hours before they decompose or photodegrade. PAHs attached to particulates may be exposed to UV light but are more resistant than those that are in the gaseous phase. Ozone can also be effective in degradation but has less of an effect than sunlight. PAHs that lie on the surface of soil may rapidly decay because they too are exposed to UV light before they are absorbed into the dirt. Unlike PCBs, dioxins, and furans, PAHs are not listed as persistent organic pollutants (POPs) by the Stockholm Convention, but some chlorinated versions have been proposed for listing.

There are so many factors that affect the stability of PAHs that assigning precise half-life times can be misleading. Rather, it is safe to say that two-ring PAHs have half-lives of days or less in the atmosphere, weeks in water, months in soils, and years in sediments. Those with three to four rings have half-lives that are about double those of two-ring structures. PAHs with five or more rings persist for weeks in the atmosphere, months in water, and several years in sediments and soil.

7.2.4 ENVIRONMENTAL CONCENTRATIONS

PAHs may be found in a wide variety of products, including coal tar, creosote, petroleum, firewood, asphalt, and some dyes, but their deliberate manufacture is limited. A few PAHs such as naphthalene are manufactured for pharmaceuticals, and others may be found in plastics.

PAHs typically come in mixtures. Seldom, if ever, do we see an area with only one or two different chemicals. A multitude of PAHs in an area, each with its own toxicity, makes it difficult to compare risk from one area to another. Sites may be comparable in their concentration of total PAH but have very different compositions and therefore pose different risks to organisms including humans.

Atmospheric concentrations are often in the low ng/m^3 (parts per trillion) range. As we have said, these lightweight molecules form a major portion of the gaseous state of PAHs, whereas the middle-weight molecules are usually associated with particulate matter in the atmosphere. Some long-term monitoring studies suggest that PAH concentrations in air are declining slowly but anthropogenic sources maintain the presence of these chemicals. In urbanized areas, PAH concentrations in air tend to be higher in winter than in summer, which makes sense if you consider that winter results in increased burning of fuels for heat. Following a similar logic, the atmosphere above cities tends to have higher concentrations than that of rural areas (Eisler, 2000). Eisler also suggested that the concentration of the carcinogen benzo[a]pyrene (BaP) provides a rough estimate of total PAHs in the air, with the total being about 10 times greater than the BaP concentration. Mean concentrations of BaP over the Great Lakes range from very low or nondetectable around Lake Superior to around 0.6 ng/m^3 over Chicago (EPA, 2014).

Worldwide, total concentrations in water tend to be at the low µg/L (parts per billion) levels, but near sources of contamination such as oil spills, concentrations may be 1000 times greater than that. The low solubility and tendency to volatize reduces the concern about PAH contamination in water.

Sediment and soil concentrations of total PAH range broadly from below detection limits in reference areas to several μg/kg (parts per billion) or even mg/kg (parts per million) in contaminated sites. Similarly, specific PAHs are going to have higher concentrations where they are used for industrial purposes. For instance, creosote and other preservatives used to maintain wood products such as railroad ties and telephone poles are high in naphthalene, phenanthrene, and pyrene, and soil concentrations near wood preserving plants are similarly high in these PAHs. Ellis et al. (1991) determined that subsoil concentrations of phenanthrene were as high as 3400 mg/kg (3.4%) near creosote plants and anthracene peaked at 693 mg/kg. In comparison, benzo[a]pyrene, chrysene, and some other common PAHs, which are not used in creosote, were absent altogether.

7.2.5 SOME EXAMPLES OF BIOLOGICAL CONCENTRATIONS

PAHs have a high tendency to bioconcentrate from the environment to organisms. This is facilitated by the lipophilic nature of PAHs; animal fats and plant oils often have the highest concentrations within an organism. Biological concentration factors (BCFs) vary among specific PAHs; the lowest and highest molecular weight contaminants are less likely to bioconcentrate than the middle-weight molecules. Other factors include the length of exposure, the physiology of the organism of concern, and the tissue. Eisler (2000) summarized BCFs from several different studies, and the values ranged widely. For example, BCFs for anthracene ranged from 200 in the water flea (*Daphnia magna*) exposed for 1 hour to 9200 in the rainbow trout (*Oncorhynchus mykiss*) exposed for 72 hours. For benzo[a]pyrene, BCFs ranged from 9 in clams (*Rangia cuneata*) exposed for 24 hours to over 134,000 in another species of water flea (*Daphnia pulex*) in a laboratory solution for 3 days.

Figure 7.2 illustrates some of the relationships between time of exposure and organ in the uptake of benzo[a]pyrene by northern pike (*Esox lucius*, Balk et al., 1984). The authors stated that most of the PAH was taken up by tissues within a day and then dispersed into organs. We can see that the BCF for benzo[a]pyrene increases early but tapers off after several days. This decrease is commonly seen as organisms adjust to the presence of PAH and activate mechanisms, including the P450 system, to increase their metabolism and depuration. Also, note that bile had the highest concentration. Bile is one way to depurate PAHs as it secretes into the intestines and passes with feces.

7.2.6 BIOLOGICAL EFFECTS OF PAHs

Terrestrial plants can absorb PAHs, especially low molecular weight ones, from the soil through their roots and move them to leaves and stems. Under natural conditions, plants do not readily bioconcentrate the contaminants and usually have lower concentrations than their immediate environment. However, soil particles can cling to leafy surfaces and the total PAHs on a plant may be as high as or higher than its surrounding soils.

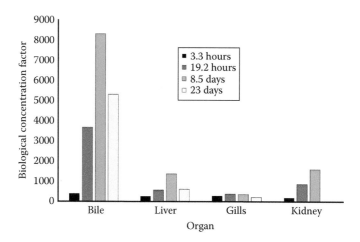

FIGURE 7.2 Biological concentration factors (BCFs) for benzo[a]pyrene in northern pike tissues based on length of exposure. BCFs have been reduced to 1/10 of actual in bile for presentation.

Toxic effects on green plants are rare at environmentally realistic concentrations. However, Dubrovskaya et al. (2014) determined that phenanthrene at 10 and 100 mg/kg interfered with several physiological processes in sorghum (*Sorghum bicolor*), including photosynthetic pigments, germination, survival of seedlings, and growth. Cruciferous plants such as kale, cabbage, Brussels sprouts, etc., contain anticarcinogenic chemicals (glucosinolates) that protect them against the toxic effects of PAHs; their protective effects can be transferred to animals, including humans that eat them (ATSDR, 2009).

In general, PAHs can cause disorders including depressed growth and survival, genotoxicity, metabolic effects, reproductive suppression, and endocrine disruption (ED). For most PAHs concentrations that cause acute lethality in the laboratory are at least one to two orders of magnitude higher than those routinely found in the environment, thereby reducing concern for acute lethal effects in the field. The lighter weight, two- and three-ring PAHs tend to have higher acute toxicity than larger molecules; four- and five-ring PAHs tend to be the most carcinogenic; and heavier PAHs tend to be relatively biologically inert, possibly because they have such low solubility (Eisler, 2000). Benzo[a]pyrene is considered to be the most carcinogenic of the saturated PAHs, but some of the methylated PAHs also have very pronounced carcinogenic activity. Thus, BaP is often used as a standard by which other saturated PAHs are compared.

Regarding cancer, the presence of a **bay region** or **fjord region** in the molecular structure of the PAH (Figure 7.3) is a useful predictor of potential PAH carcinogenicity. This type of structure facilitates linkage between PAHs and sections of the DNA molecule. We know, for instance, that the bay region of BaP allows it to bind with guanine of DNA and RNA. Once attached, BaP can disrupt protein synthesis, either preventing a required enzyme from being processed or causing harmful alterations in others.

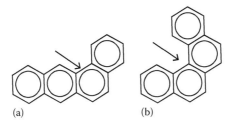

FIGURE 7.3 PAHs with a 'bay' or 'fjord' region in their structures have a greater propensity to cause cancer than those molecules that do not have these regions. (a) Bay structure and (b) fjord structure. (Diagram from Muñoz and Albores, 2011.)

The Cytochrome P450 system also functions in the metabolism of PAH, as it does in other organic contaminates that we have discussed. Through a set of complex reactions, PAHs enter different metabolic pathways that can produce carcinogenic derivatives. The parent molecules are not toxic in themselves but must be converted by the P450 system to become toxic. One metabolic route is for the PAH to become oxidized and the bay or fjord region makes oxidation easier. Oxidation can then lead to chemicals that enter the nucleus of cells where they can bind with DNA to form adducts (i.e., attachments, Figure 7.4). These adducts interfere with normal translation of proteins and are particularly risky if they take place in rapidly growing or replicating tissues such as bone marrow, or skin.

A great many studies have been conducted on the effects of PAH on fish. As we discussed in Chapter 3, brown bullheads (*Ameiurus nebulosus*) have been taken from the Detroit River, the Anacostia River in Washington, DC, the South River in Maryland, and the Black River in Ohio with many maladies, including truncated barbels, skin lesions, and liver lesions. The frequency of occurrence of these problems correlated with the concentration of PAHs in sediments in Detroit (Leadley et al., 1998) and the Black River (Baumann and Harshbarger, 1995).

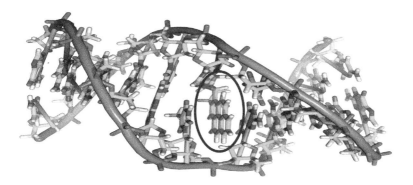

FIGURE 7.4 Illustration of a metabolite of BaP (circled) forming an adduct with DNA. (Courtesy of Wikipedia, https://en.wikipedia.org/wiki/DNA_adduct.)

Some research has been done on the effects of PAHs on amphibians. In one study, tiger salamanders (*Ambystoma tigrinum*) were captured from a contaminated pond on a Texas Air Force base (Anderson et al., 1982). Several PAH-containing pollutants had either been dumped or allowed to run off into the pond over many years, and the salamanders had a 24%–42% rate of skin tumors, some of which were benign, and others cancerous.

Bommarito et al. (2010) studied the effects of two road sealants, coal tar and asphalt, on eastern newts (*Notophthalmus viridescens*). Sediments containing the sealants from 0 to 1500 mg/kg were placed in aquaria under either UV radiation or visible light to determine concentration/response relationships. The total PAH in the asphalt treatment ranged from 0.07 to 20.58 mg/kg in sediment and 30 to 281 µg/L in water. For coal tar sediment, concentrations ran from 1.51 to 1149 mg/kg and for water from 30 to 1464 µg/L. The concentrations of PAHs were not lethal after 28 days, but significant effects due to the sealants included decreased righting ability and diminished liver enzyme activities. Coal tar sealant was more effective in inducing these changes than asphalt sealant. In neither case was serious damage observed, but it would have been interesting if the animals could have been kept for several months to determine if any tumors developed.

Many studies on the effects of PAHs on birds have been associated with oil spills because shorebirds are among the fauna that are often heavily affected by such spills. For example, Bustanes (2013) related where common eiders (*Somateria mollissima*) in Norway suffered population declines during the 1980s when PAH pollution was being dumped into surrounding waters. The population declined from 800 breeding pairs to less than 500 pairs over several years. Hatching success in the affected area was around 90% compared to nearly 100% in a reference area. Apparently, blue mussels (*Mytilus edulis*), a primary food for the eiders, had accumulated very high concentrations of PAH and had BCFs around 500. Eiders would eat the mussels and pass the PAHs to their eggs through maternal transfer. After oil leakages were stopped in the mid-1990s, the population began to return, and by 2000 there were again 700 breeding pairs of eiders. Oil has many toxic components besides PAH, and Bustanes admitted that the evidence for PAH causing the decline was circumstantial, but that's often what happens in biological field studies—it is often difficult to identify a specific contaminant for population declines. In this study, however, the only group of contaminants that was apparent after extensive sampling was PAH.

7.3 OIL SPILLS AND PAH

Do you have any idea how many oil spills occur each year in the United States? If you guessed 1,000, 5,000, or even 10,000, you would be way off the mark. The U.S. EPA (2016b) estimates that upwards of 14,000 spills containing 10–25 million gallons (38–94 million liters) occur each year in this country alone. Add to that the number of spills in other countries such as the Middle East, and you have a very substantial amount of toxic oil dumped into the environment. We do seem to be getting better over time in conserving our oil, at least in terms of tanker spills, but there are events such as the BP Deepwater Horizon oil spill that continue to raise caution.

The top 10 oil spill disasters in history resulted in conservative estimates of a total of 30–34 million barrels of oil (5.74–6.38 billion liters). The BP oil spill, the largest in U.S. history, resulted in 3.3–5 million barrels of oil dumped into the Gulf of Mexico (Smithsonian, 2014).

Crude oil or petroleum is a complex of thousands of hydrocarbon and nonhydrocarbon substances (Albers, 2003). By weight, most crude oil and refined oils are composed of 75% hydrocarbons, but heavy crude can contain upwards of 50% nonhydrocarbons. Crude oil contains 0.2%–7% PAH, with the percent PAH increasing with the specific gravity of the oil.

Petroleum damages vegetation in many ways. Mangroves have been frequently hard hit by oil pollution in tropical and semitropical regions because mangrove trees grow very thickly and trap oil among their trunks and root systems. Decaying oil sinks to the bottom and can decrease the already limited oxygen supply to the roots. Other plant communities that are often affected include salt marshes, large intertidal algae beds, sea grass communities, and freshwater vegetation. Harmful effects of petroleum to plants often involve coating of leaves, thereby reducing photosynthesis and growth.

While freshwater and marine fish are usually very mobile and can escape major effects of oil spills, if spills occur in small, disconnected bodies of water or rapidly move through the water, they can affect fish populations. Eggs, however, are not mobile and larvae may be less able to escape than adults, so the smothering effect of petroleum can quickly kill embryos and larvae. Thus, timing can be very important in determining the overall effect of spills; spills that occur during breeding seasons are likely to be the most severe.

Very little is known about the effects of oil spills on amphibians or reptiles. Sea snakes may be smothered and sea turtles may consume toxic levels of oil. Following the BP oil spill, 457 sea turtles were found with visible oiling. Of these, 18 were dead and 439 were alive. An additional 517 turtles were found dead by search crews with no visible oiling and 80 live turtles that were collected had no oil (FWS, 2011). More recent studies suggest that the impacts of the oil spill may have affected sea turtle populations across the Atlantic (Putman et al., 2015).

Seabirds and marine mammals are among the hardest hit of animals from spills. The Exxon Valdez crude oil spill created a 750 km slick and killed an estimated 250,000 seabirds, 2,800 sea otters (*Enhydra lutris*), and 22 orca (*Orcinus orca*, Malakoff, 2014). The embryos of birds are particularly sensitive to oil pollution. Oil-contaminated hens return to the nest and drip oil on eggs. As little as 1–20 μL of light oil that penetrates the eggshell can cause death (Hoffman, 1990). Since a mL of fluid is roughly equivalent to 10 drops and a μL is 1/1000 of a mL, much less than a drop can result in dead embryos.

7.4   ORGANOCHLORINE PESTICIDES

These are molecules that have at least one phenol group and multiple chlorine atoms. They were synthesized to combat pests, insects, weeds, or some other undesirable organisms. As a family of chemicals, OCPs are used to manufacture vinyl chloride, much of which goes into PVC plastics or is used as insulating

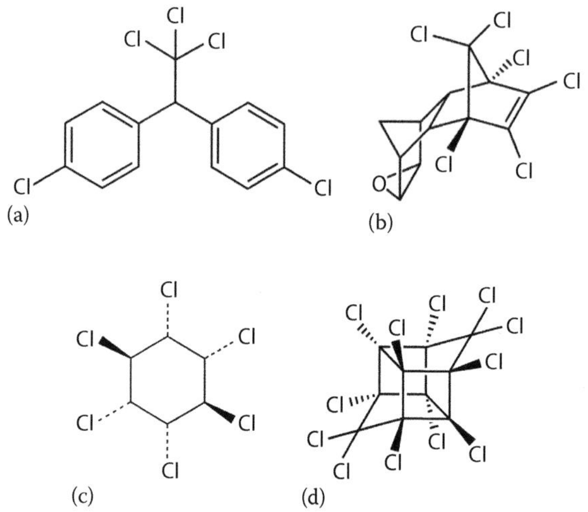

FIGURE 7.5 Examples of organochlorine pesticides (OCPs): (a) p,p′-DDT; (b) endrin; (c) lindane; and (d) mirex. In these diagrams, dark lines denote that the atom is projecting toward the viewer, while those that are dashed indicate that the atom goes away from the viewer.

materials and solvents. However, their most notorious use has been as pesticides. The pesticides include many different compounds including dichlorodiphenyltrichloroethane (DDT) and its relatives, lindane (gamma-hexachlorocyclohexane), methoxychlor, endrin, endosulfan, heptachlor, chlordane, toxaphene, and mirex, to list just a few (see Figure 7.5 for some of these compounds). Over the years, all but a few organochlorines (OCs) have been banned in the United States, Canada, the European Union, and elsewhere. A couple of OCs are still used in veterinary formulations in the United States to repel fleas and ticks, lindane is still used to combat head lice and scabies in humans, and the use of DDT has been allowed to continue in malaria-plagued areas to reduce mosquito populations. Despite their discontinued use, many OCs are still to be found in the environment due to their persistence.

The intended function of most OCPs was to control insects by interfering with their nervous systems. Specifically, DDT and related compounds operate on sodium, potassium, and iodine channels along axons of neurons, so that the rate of action potentials is increased uncontrollably. This causes uncontrolled, repetitive, spontaneous discharges along the axons of neurons, which can lead to loss of control in the muscles used for respiration, and animals will usually die from asphyxiation. A subclass of OCPs called cyclodienes, which includes endrin, aldrin, endosulfan, and others, acts on the gaba amino-butyric acid (GABA) receptors of the neuron to produce similar results. Typical symptoms in animals exposed to these pesticides include convulsions, agitation, excitability, and other nervous disorders leading to acute lethal toxicity. In addition, chronic, sublethal effects of OCPs include endocrine disruption, cancer, disruption of enzyme functions, and numerous other physiological disorders.

Most OCPs are lipophilic and can be stored in fat, thus promoting bioconcentration and biomagnification. Most are very persistent compounds in the environment, lasting years or decades after application, and thus have been identified as POPs by the Stockholm Convention. Sometimes, they are also referred to as legacy compounds because they are still found in the environment years and even decades after their use was discontinued.

7.4.1 Sources and Use

OCPs became popular as insecticides in the 1930s. DDT was initially synthesized in 1874 but wasn't used as a pesticide until 1939. At that time, it was considered a boon because it was a cheap, effective way of reducing insect pests on crops. Through the 1960s, other OCPs were manufactured in the United States. However, as the use of these OCPs continued, insects began developing resistance and effective concentrations had to be increased every few years. Based primarily on human health risks and their persistence, the EPA banned DDT in 1972, Canada did so a few years later, and Mexico banned it in 2000. However, the use of DDT is allowed in a limited context for malaria control. There does not seem to be any pesticide that is as cheap and effective in controlling the mosquito vectors of malaria as DDT. In most cases, DDT is sprayed on the walls of houses and huts and cannot legally be applied to wetlands. Unfortunately, the mosquitoes in these countries are developing the same resistance to DDT as was seen in crop pests in the United States prior to it being banned. Additionally, India remains highly contaminated by DDT and hexachlorocyclohexanes (HCHs), in part due to the increasing use of OCPs in mosquito control (Sharma et al., 2014).

Banning means that the compound is no longer approved by the responsible government agency and can no longer be used. In 2010, the EPA began a phaseout of endosulfan, the most recent OCP to be restricted.

7.4.2 General Chemical Characteristics of OCPs

Table 7.2 presents some general chemical characteristics of OCPs. A notable feature is low solubility in water and a high lipophilicity shown by the relatively high log K_{ow}s of many OCPs. Lindane, pentachlorophenol, and chlorobenzilate have relatively high water solubility and may be found dissolved in water at toxic concentrations. Mirex, DDT, and aldrin with lower water solubility would be less likely to be in the water column of lakes, ponds, or rivers. Mirex and DDE readily bind with soil and sediments, another reason that they are unlikely to be found free in water. The melting points of all the listed OCPs show that they are solids at room temperature, and most are white or colorless crystalline structures.

7.4.3 Structure

Structurally, OCPs fall into five classes (Figure 7.6): (1) DDT and its analogs including DDE; (2) HCHs such as lindane; (3) cyclodienes including aldrin, dieldrin, endrin, heptachlor, chlordane, and endosulfan; (4) toxaphene; and (5) mirex

TABLE 7.2

Some Physical Properties of Selected Organochlorine Pesticides

Common Name	MW (g/mol)	Melting Point (°C)	Water Solubility (mg/L)	K_{ow}	Vapor Pressure (mmHg)	Approximate Half-Life in Field	U.S. Status (Year Banned)
Aldrin	365	104	0.03	6.5	7.5×10^{-5}	345 days	1974
Chlordane	410	104	0.1	6.0	8.7×10^{-6}	3.8 years	1988
DDE	318	88.4	0.06	6.9	6×10^{-8}	15+ years	1972
DDT	355	227	0.001	6.0	1.6×10^{-7}	15+ years	1972
Dicofol	370	78.5	0.8	4.3	5.9×10^{-8}	16 days	Miticide
Dieldrin	381	176	0.19	3.6	4×10^{-7}	2.7 years	1987
Endosulfan	407	106	0.5	3.1	1.7×10^{-7}	200 days	Phasing out
Heptachlor	373	95	0.06	4.4	4×10^{-4}	250 days	Fire ant control

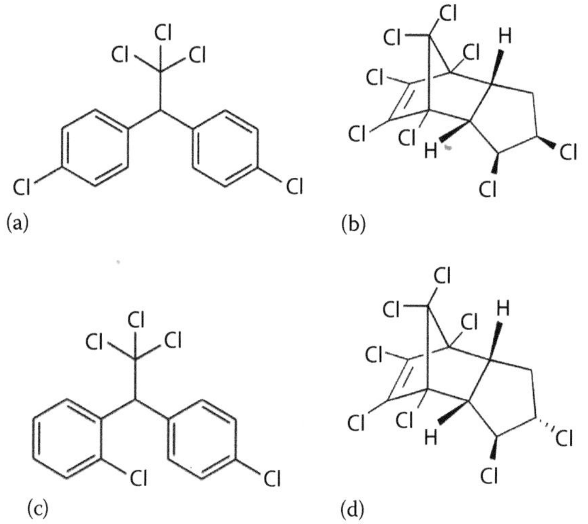

FIGURE 7.6 Examples of some OCPs: (a) p,p′-DDT; (b) *cis*-chlordane; (c) o,p-DDT; (d) *trans*-chlordane. p,p′-DDT and o,p′-DDT are enantiomers of each other and *cis*- and *trans*-chlordane are diastereomers.

and chlordecone. Toxaphene is made up of approximately 200 chemicals, many of which are proprietary and not identified by the manufacturer. Technical grade chlordane is also a mixture of over 120 structurally related chemical compounds, but it is primarily composed of different forms of chlordane. Many OCPs are represented by different stereoisomers that have the same molecules and molecular weights but differ slightly in arrangement (Figure 7.6). These stereoisomers can differ radically in their chemical and biological properties. For instance, the DDT family includes DDT, dichlorodiphenyldichloroethane (DDD), and dichloro-2,2-bis(p-chlorophenyl) ethylene (DDE), and all have ortho- and para-isomers. Technically, these are called enantiomers and DDD and DDE are metabolites of DDT. Of these, p,p′-DDE is arguably the most toxic, although each has its own toxic characteristics. In contrast, when you see *cis-* and *trans-* as in *cis*-chlordane and *trans*-chlordane or the chemical name is preceded by a Greek letter as in α-endosulfan or β-endosulfan, we are discussing diastereomers. Diastereomers can be transformed into each other during metabolism, whereas enantiomers cannot.

7.4.4 Persistence

While Table 7.2 lists approximate half-lives of the compounds, actual persistence in the environment is affected by many different factors. Although they are resistant to biodegrading, some strains of bacteria, fungi, and invertebrates have the ability to slowly break down the parent compounds. Also, certain plants can metabolically degrade OCPs. Physical factors that help reduce OCPs include sunlight, pH, and moisture. Low pH and the presence of ions such as CO_3^- and HCO_3^- accelerate degradation. Ultraviolet radiation is very significant in degrading OCPs. Eventually, the factors of degradation cause even the most stable POPs to break down.

Persistence also varies by chemicals. While DDT is somewhat stable in the environment, it is degraded into DDE, which is even more persistent. For that reason, residue studies often find higher DDE than DDT concentrations. In organisms, the ratio of DDE to DDT can be even higher because some conversion takes place through the P450 system. Heptachlor and aldrin are rapidly metabolized by the body and have short half-lives in organisms. However, their metabolites may be just as or sometimes even more toxic than the parent compound. Heptachlor epoxide, the major breakdown product of heptachlor, is far more stable in the body and the environment; both forms have about the same toxicity (ATSDR, 2009). Aldrin breaks down into dieldrin and its stereoisomer, endrin. Aldrin is not toxic to insects (Blus, 2003), but endrin has one of the highest toxicities among OCPs.

As an example of OCP decline in wildlife, let us use the peregrine falcon (*Falco peregrinus*) at San Padre Island, Texas, from 1978 to 2004 (Henny et al., 2009). p,p′-DDT dropped from 0.879 μg/g wet weight in 1978 to 0.013 μg/g in 2004, a decrease of 98%. Moreover, the frequency of occurrence for several OCPs including p,p′-DDT, p,p′-DDE, heptachlor epoxide, dieldrin, *cis*-chlordane, and mirex went from 10%–43% of the birds sampled in 1978 to zero in 2004. So it looks like, given enough time, we may eventually see the end of these legacy compounds.

Lipophilic OCPs have a tendency to biomagnify. One of the great concerns about DDT was that it could start in parts per billion in water, for example, but by the time the top carnivore level was reached, concentrations were millions of times greater. In the environment you can start with the DDT attached to organic particulates in water and animals such as zooplankton assimilate these OCPs. These microorganisms are ingested by aquatic insects, which are consumed by small fish that are ingested by bigger fish, etc., and at each step the concentrations of OCPs increase (Figure 7.7). Because initial concentrations of OCPs are very low, effects are not observed until the highest trophic levels are reached. This process of biomagnification was very important in the population declines seen in many fish-eating raptors such as brown

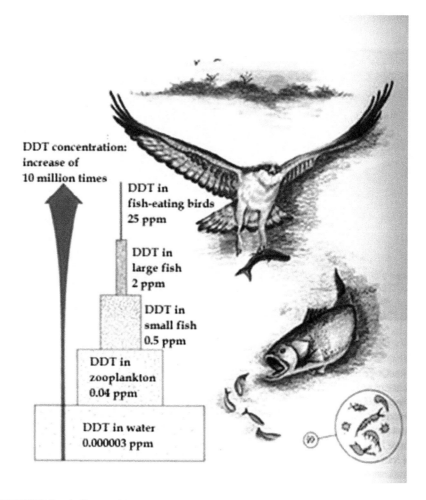

FIGURE 7.7 A diagram illustrating how DDT biomagnifies up the food chain. Concentrations in water as little as 3×10^{-6} parts per million can reach 10 million times than concentration in top predators. (Courtesy of U.S. Fish and Wildlife Service, https://usresponserestoration. wordpress.com/tag/pesticides/.)

pelicans (*Pelecanus occidentalis*), bald eagles (*Haliaeetus leucocephalus*), and osprey (*Pandion haliaetus*) in the 1980s and 1990s.

7.4.5 EXAMPLES OF OCP CONCENTRATIONS IN ENVIRONMENTAL SOURCES

Initially, OCPs were introduced into the environment as modern pesticide application today through spraying over agricultural fields (Figure 7.8). Spraying over wetlands (intentional or accidental), aerial transport, spills, leaching of stockpiles or dump sites, and direct discharge of effluents or emissions from industrial processes were access points to the environment. Back in the 1960s and 1970s, dump sites

(a)   (b)

FIGURE 7.8   Two ways in which pesticides are commonly applied to agricultural fields: (a) through ground spraying and (b) aerial spraying. In both cases, fine droplets of pesticide can be carried for many miles by winds.

and landfills did not receive the maintenance that they do today. From there, these persistent contaminants could be recycled through various parts of the environment. If discharged into water, they are most likely to adhere to sediments. Sediment-bound OCPs usually are covered over through time. OCPs can volatilize from dry soil, which is one way that they are released into the atmosphere in areas where they are no longer used but still exist. Eventually, OCPs can return to water or soil as wet or dry precipitation. Weather factors such as snowmelt or rain can cause runoff into streams or wetlands, whereas wind erosion can resuspend the contaminants in the atmosphere.

Of the four major abiotic compartments in the environment—air, water, sediments, or soil—sediments generally contain the highest concentrations of OCPs, followed by soil; water and air have low but often measureable concentrations. Units of measurement for tissues, soil, or sediment are usually expressed as ng/g or μg/kg. In water, the concentrations are usually much lower and measured as ng/L. In air, the concentrations are even lower and measured as pg/m^3.

7.4.6   CONCENTRATIONS OF OCPs IN ANIMALS

Concentrations of OCPs in organisms can differ by three orders of magnitude or more across animal samples depending on species, age, sex, organ, food, and several environmental factors. After species and tissues, location is a key factor in predicting OCP concentrations in organisms. If an animal or plant is in or near a contaminated site, obviously they may be expected to have higher concentrations of contaminants than those from reference or clean sites. For sedentary species, this means that the OCPs have to be in their home range. Migratory species can take in OCPs from their winter or their summer ranges, or both. For the most part, terrestrial species ingest OCPs with prey organisms or ingested contaminated soil. Aquatic organisms such as freshwater and marine fish as well as marine mammals such as whales and dolphins may be constantly bathed in contaminated waters. Many of these animals also ingest contaminated prey.

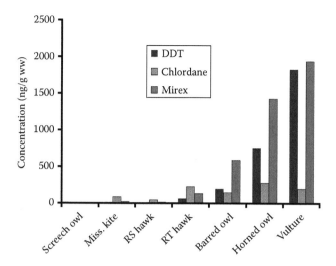

FIGURE 7.9   Mean concentrations of OCPs measured in livers of raptors collected from coastal South Carolina, 2009–2010. Levels of chlorinated, brominated, and perfluorinated contaminants in birds of prey spanning multiple trophic levels. Miss., Mississippi; RS, red-shouldered; RT, red-tailed; vulture, turkey vulture. (Adapted from Yordy, J.E. et al., *J. Wildl. Dis.*, 49, 347, 2013.)

Diet is another important factor that can affect the concentration of OCPs, but this is often closely associated with focal species. For example, Yordy et al. (2013) found that birds of prey collected in the same area could vary considerably in their OCP concentrations (Figure 7.9). There was an overlap in many of the diets, but still we see a big difference between the screech owl (*Asio otus*) that feeds on insects and small rodents and the turkey vulture (*Cathartes aura*) that is a scavenger. One great horned owl (*Bubo virginianus*, not reported in the figure due to its very high contaminant levels) had hundreds of times the concentration of DDT than many of the other species, 10 times higher than even the turkey vulture.

After location and diet, other factors play case-specific roles in affecting contaminant concentrations. Age is sometimes important because older animals may have accumulated more fat than very young animals and, through time, there can be a greater bioconcentration of lipid-soluble contaminants. Sex can also be a factor if one tends to have more body fat than the other or if females can off-load PCBs through maternal transfer. As an example, blubber of adult male common bottlenose dolphins (*Tursiops truncatus*) living off the coast of Sarasota, Florida, had significantly higher concentrations of several OCPs than juvenile males or females and 10–40 times the amount found in adult females (Yordy et al., 2010). The dolphins were in the process of or recently completed weaning the young and had recommenced eating fish, which contributed to their contaminant body burdens. The authors attributed the lower concentrations of OCPs in the adult females to their ability to off-load some OCPs into their milk.

OCPs are not equally distributed through an organism. In most cases, organs and tissues with greater lipids will have higher OCP concentrations than the lean compartments. For this reason, body concentrations of OCPs, like some other organic contaminants, are often measured in terms of percent lipids. Lipid-adjusted measurements standardize concentrations from one tissue to another and among animals.

For example, in the Patagonian silverside (*Odontesthes hatcheri*), a South American fish, p,p'-DDE had the highest concentrations among the DDTs and, based on concentrations in lipids, had the following sequence in tissues: liver > gill > muscle > gonads (Ondarza et al., 2014). Because the silverside is commonly eaten by humans, the authors wanted to determine if eating the fish could be harmful. Its muscle has the highest rate of consumption and the authors decided that the concentrations of OCPs were not of concern; however, PCBs had higher concentrations than OCs and were considered potentially harmful to human health.

7.4.7 BIOLOGICAL EFFECTS OF ORGANOCHLORINE PESTICIDES

Relatively few studies have been conducted on the effects of OCPs on plants. Some studies have uncovered chromosomal damage in nonreproductive tissues, others have detected reduced photosynthesis and growth, but most of these have tested concentrations higher than what would be expected in nature.

OCPs can cause a diverse set of effects on animals. The most infamous of these effects include cancer and endocrine disruption evidenced in all vertebrates. However, many other sublethal effects due to OCPs have been reported in an extensive literature. The focus has been on sublethal effects because the amount of most OCPs that would produce acute lethality is higher than what would be expected in the environment. Death due to OCP toxicity is often slow and occurs as a general wasting away. The tendency for OCPs to be stored in fat can lead to a re-release of the contaminants during food shortages when body reserves are used. This can cause delays from the time of first exposure to the onset of effects. DDT can remain in the human body for 50 years or more (Mrema et al., 2013), and we have to assume that other animals can store OCPs similarly, only to be released later on.

As with virtually every contaminant we have seen, OCPs are highly toxic to aquatic organisms but less toxic to birds or mammals. For example, chlordane is lethal at 24 µg/L to copepods, 13 µg/L to goldfish (*Carassius auratus*), and 3 µg/L to largemouth bass (*Micropterus salmoides*). But diets a 1000 times higher in concentration had no effect on several species of birds and mammals. Even at that, however, OCPs have been directly involved in the deaths of wild animals, including fish and birds. During a 16-month period in 1996 and 1997, Stansley and Roscoe (1999) documented chlordane poisoning in six species of songbirds and four species of raptors in New Jersey. At one roost, they recovered 425 dead or sick birds, including common grackles (*Quiscalus quiscula*), European starlings (*Sturnus vulgaris*), and American robins (*Turdus migratorius*). Sick birds displayed signs that were consistent with chlordane poisoning, including convulsions, muscle spasms, and excessive vocalizing. Chlordane poisoning was also diagnosed in Cooper's hawk (*Accipiter cooperii*). The timing of Cooper's hawk mortalities coincided loosely with the July peak in songbird mortalities and may have been due to hawks feeding on other birds debilitated by chlordane.

Aldrin, which has very low toxicity by itself, is readily converted to dieldrin or endrin in the field and in organisms, and both of these are very toxic to birds and mammals. An early example of wildlife mortality due to aldrin/dieldrin occurred in Illinois when aldrin was applied on farm fields to control Japanese beetles (*Popillia japonica*, Labisky and Lutz, 1967). The application was effective on the beetles, but it also wiped out 25%–50% of the pheasants living in sprayed areas and severely depressed reproduction; more than half of the pheasant hens lost broods that year. Fortunately, death and diminished reproduction only lasted the year of application.

Endrin was probably the chief OCP responsible for the extirpation of brown pelicans in Louisiana during the early 1960s. A large discharge of endrin into the Mississippi River from a plant in Tennessee resulted in the chemical moving downstream toward the Gulf of Mexico. Along the way, millions of fish perished with mortality increasing as the pesticide approached the Gulf. Gulf menhaden (*Brevoortia patronus*), a baitfish, took one of the biggest hits. Various estimates of the preexposed Louisiana population of brown pelicans at this time ranged from 12,000 to 85,000 birds, but the entire population was extirpated. Based on endrin residues and reproductive declines in the pelicans, Blus et al. (1979) made a convincing argument that endrin was responsible for the extirpation of this species through the decrease in fish, with toxicity to the birds as a secondary factor.

OCPs disrupt certain hormones, enzymes, growth factors, and neurotransmitters and induce key genes involved in the metabolism of steroids and xenobiotics (substances that are foreign to the body). These disruptions cause complex and not completely understood changes in many different parts of the body. A direct of these influences is an alteration in the homeostatic condition of cells, leading to oxidative stress and accelerated **apoptosis** or cell death. In healthy individuals, apoptosis is a way of removing damaged or unnecessary cells without causing damage to surrounding cells. The abnormal apoptosis caused by OCPs is related to various pathologies including immunodeficiency, autoimmune diseases, cancer, and reproductive problems (Mrema et al., 2013).

OCPs also exert a multitude of genetic effects involving DNA functioning and gene transcription. These gene alterations can result in permanent changes in developmental patterns, leading to endocrine disruption and reduced immunosuppression. This is an area of active research with many unanswered questions, but Mrema et al. (2013) expressed concern that most of the related studies are using environmentally unrealistic concentrations, thereby confounding what might be occurring in nature.

The endocrine disruption effect of OCPs occurs through the contaminants interfering with the endocrine system at several points. One method is at those cells that typically respond to hormones. Hormones attach to specialized binding receptors on these cells and exert their effects. In the case of reproduction, cells in the ovary, testes, or other related organs possess many of these receptors, while those in other organs such as the heart have few or maybe even none. DDT and its derivatives, endosulfan, and lindane mimic the effects of hormones by binding to these receptors and triggering their effects. Other OCPs interfere with various enzyme pathways responsible for the synthesis of hormones. Thus, DDT, particularly o,p′-DDT, mimics natural estrogens (Petersen and Tollefsen, 2011). In contrast, p,p′-DDE and other OCPs can act as antiandrogens, counteracting the effects of testosterone (Misaki et al., 2015). Other studies have shown that OCPs can interfere with normal thyroid functioning (Brown et al., 2004).

A lot can be learned about the real and potential effects of OCPs or other contaminants using controlled laboratory studies at concentrations that might not be expected to occur under natural conditions. It is usually very difficult to determine if a specific OCP or any group of contaminants is actually having an effect on free-ranging wildlife. In addition to the chemicals of concern, if animals are living in a contaminated environment, they typically are exposed to many different chemicals simultaneously. In addition, there are other sources of stress such as disease, predation, and competition that are also working on individuals and their populations, and these can mask the effects of the contaminants.

7.5 CHAPTER SUMMARY

1. Most polycyclic aromatic hydrocarbons (PAHs) are composed of two or more phenol rings, with carbon and hydrogen as the only molecules. Others, called substituted PAHs, have other organic molecules, amines, hydroxyls, or oxygens attached.

2. PAHs are naturally produced during the process of combustion. As a result, forest fires, volcanoes, brush fires, and other sources of natural combustion are important sources of the chemicals. Man-made PAHs similarly come from combustion such as incinerator of refuse or various industrial processes.

3. PAHs vary in chemical behavior largely due to the number of rings and the molecular weight of the compound. Lightweight, one- or two-ring PAHs tend to be more water soluble and volatize more readily; heavier compounds tend to be less mobile and adhere strongly to particulate matter; middle-weight PAHs are the most likely to be in organisms.

4. Compared to other organic contaminants that we have discussed, PAHs are short-lived, having half-lives measured in days to months, depending on environmental conditions and whether the chemicals are in air, water, or soil/sediment. Many biological organisms can metabolize these chemicals, so bioconcentration is less of a concern than with some other contaminants.

5. Because of the low volatility and solubility of most PAHs, their concentrations in air and water are much lower than that in soil or sediments.

6. PAHs cause a plethora of problems to animals. However, cancer ranks as one, if not the top, concern from this chemical family. Among the PAHs, benzo[a]pyrene is the most carcinogenic of saturated forms, although some of the substituted PAHs are also very potent.

7. Thousands of oil spills occur around the world annually. Oil has direct toxicity to plants and animals but can also cause indirect lethality through smothering. A fraction of oil consists of PAHs, but other organic molecules form the majority of oil and tars.

8. Organochlorine pesticides (OCPs) include many different chemicals, all chlorinated, persistent, and lipophilic, that were the first synthetic chemical pesticides produced.

9. The more lipophilic OCPs biomagnify up food chains. They can start at very low environmental concentrations but increase until they cause severe problems in top carnivores. DDT and its analogs were responsible for the decline of several species of fish-eating birds.

10. OCPs cause myriad effects in animals. Cancer, genotoxicity, malformations, and several other illnesses have been identified in wildlife, laboratory animals, and humans.

7.6   SELF-TEST

1. True or False. Some polycyclic aromatic hydrocarbons can combine with other molecules such as amines, hydroxyls, and chlorine.

2. Among the PAHs, which of the following characteristics pertains to lightweight, one- or two-ring forms compared to larger PAHs?
   a. More water soluble
   b. Higher tendency to volatize
   c. Less chemically stable
   d. All of the above

3. The principal human source of PAHs to the environment is
   a. High temperature combustion of organic materials
   b. Oil spills
   c. Intentional industrial production
   d. Metabolic by-products

4. The most carcinogenic, saturated PAH to rats and presumably to humans is
   a. Methyl-anthracene
   b. Coronene
   c. Benzo[a]pyrene
   d. Fluoranthene

5. Short answer. In your own words, describe what is meant by a DNA adduct. What problems can occur through these adducts?

6. Short answer. Under what conditions is DDT still being used?

7. What is the intended mechanism of action for organochlorine insecticides?
   a. To block axon potentials from occurring along neurons
   b. To cause uncoordinated and repetitive firing of neurons
   c. To cause cancer in insects
   d. To sterilize male insects

8. True or False. One of the problems with using DDT is that it easily dissolves in water and thus can be easily taken in by small fish.

9. Which of the following species was severely impacted by organochlorines during the late 1960s?
   a. Brown pelicans
   b. American toads
   c. Rainbow trout
   d. Turkey vultures

10. OCPs can cause
   a. Congestive heart failure
   b. Endocrine disruption
   c. Cancer
   d. More than one but not all above

REFERENCES

Albers, P.H. 2003. Petroleum and individual polycyclic aromatic hydrocarbons. Pp. 341–371. In: Hoffman, D.J., Rattner, B.A., Burton Jr., G.A., Cairns Jr., J. (eds.), *Handbook of Ecotoxicology*, 2nd Edition. Boca Raton, FL: Lewis Publishers.

Anderson, R.S., Doos, J.E., Rose, F.L. 1982. Differential ability of *Ambystoma tigrinum* hepatic microsomes to produce mutagenic metabolites from polycyclic aromatic hydrocarbons and aromatic amines. *Cancer Lett* 16: 33–41.

Agency for Toxic Substances and Disease Registry (ATSDR). 2009. Studies in environmental medicine: Toxicity of polycyclic aromatic hydrocarbons (PAHs). http://www.atsdr.cdc.gov/csem/pah/docs/pah.pdf. Accessed December 12, 2016.

Agency for Toxic Substances and Disease Registry (ATSDR). 2016. Toxic substances. https://www.atsdr.cdc.gov/toxicsubstances.html. Accessed December 12, 2016.

Balk, L., Meijer, J., DePierre, J.W., Appelgren, L. 1984. The uptake and distribution of [3H] benzo[a]pyrene in the northern pike (*Esox lucius*). Examination by whole-body autoradiography and scintillation counting. *Toxicol Appl Pharmacol* 74: 430–449.

Baumann, P.C., Harshbargar, J.C. 1995. Decline in liver neoplasms in wild brown bullhead catfish after coking plant closes and environmental PAHs plummet. *Environ Health Perspect* 103: 168–170.

Blus, L., Cromartie, E., McNease, L., Joanen, T. 1979. Brown pelican–population status, reproductive success, and organochlorine residues in Louisiana, 1971–1976. *Bull Environ Comtam Toxicol* 22: 128–135.

Blus, L.J. 2003. Organochlorine pesticides. Pp. 313–340. In: Hoffman, D.J., Rattner, B.A., Burton Jr., G.A., Cairns Jr., J. (eds.), *Handbook of Ecotoxicology*, 2nd Edition. Boca Raton, FL: CRC Press.

Bommarito, T., Sparling, D.W., Halbrook, R.S. 2010. Toxicity of coal–tar and asphalt sealants to eastern newts, *Notophthalmus viridescens*. *Chemosphere* 81: 187–193.

Brown, S.B., Adams, B.A., Cyr, D.G., Eales, J.G. 2004. Contaminant effects on the teleost fish thyroid. *Environ Toxicol Chem* 23: 1680–1701.

Bustanes, J.O. 2013. Reproductive recovery of a common eider *Somateria mollissima* population following reductions in discharges of polycyclic aromatic hydrocarbons (PAHs). *Bull Environ Contam Toxicol* 91: 202–207.

Dubrovskaya, E.V., Polikarpova, I.O., Muratova, A.Y., Pozdnyakova, N.N., Chernyshova, M.P., Turkovskaya, O.V. 2014. Changes in physiological, biochemical, and growth parameters of sorghum in the presence of phenanthrene. *Russ J Plant Physiol* 61: 529–536.

Eisler, R. 2000. *Handbook of Chemical Risk Assessment: Health Hazards to Humans, Plants and Animals*, Vol. 2, Organics. Boca Raton, FL: Lewis Publishers.

Ellis, B., Harold, P., Kronberg, H. 1991. Bioremediation of a creosote contaminated site. *Environ Technol* 12: 447–459.

Henny, C.J., Yates, M.A., Seegar, W.S. 2009. Dramatic declines of DDE and other organochlorines in spring migrant peregrine falcons from Padre Island, Texas, 1978–2004. *J Raptor Res* 43: 37–42.

Hoffman, D.J. 1990. Embryotoxicity and teratogenicity of environmental contaminants to bird eggs. *Rev Environ Contam Toxicol* 115: 39–45.

Labisky, R.F., Lutz, R.W. 1967. Responses of wild pheasants to solid-block applications of aldrin. *J Wildl Manage* 31: 13–24.

Leadley, T.A., Balch, G., Metcalfe, C.D., Lazar, R., Mazak, E. et al. 1998. Chemical accumulation and toxicological stress in three brown bullhead (*Ameiurus nebulosus*) populations of the Detroit River, Michigan, USA. *Environ Toxicol Chem* 17: 1756–1766.

Malakoff, D. 2014. 25 years after the *Exxon Valdez*, where are the herring? *Science* 343: 1416.

Misaki, K., Suzuki, G., Tue, N.M., Takahashi, S., Someva, M.H. et al. 2015. Toxic identification and evaluation of androgen receptor antagonistic activities in acid-treated liver extracts of high-trophic level wild animals from Japan. *Environ Sci Technol* 49: 11840–11848.

Mrema, E.J., Rubino, F.M., Brambilla, G., Moretto, A., Tssatsakis, A.M., Colosio, C. 2013. Persistent organochlorinated pesticides and mechanisms of their toxicity. *Toxicology* 307: 74–88.

Muñoz, B., Albores, A. 2011. DNA damage caused by polycyclic aromatic hydrocarbons: Mechanisms and markers, In: Chen, C. (ed.), *Selected Topics in DNA Repair*, InTech, ISBN: 978-953-307-606-5. http://www.intechopen.com/books/selected-topics-in-dna-repair/dna-damage-causedby-polycyclic-aromatic-hydrocarbons-mechanisms-and-markers.

Ondarza, P.M., Gonzalez, M., Fillmann, G., Miglioranza, K.S.B. 2014. PBDEs, PCBs and organochlorine pesticides distribution in edible fish from Negro River basin, Argentinean Patagonia. *Chemosphere* 94: 135–142.

Petersen, K., Tollefesen, K.E. 2011. Assessing combined toxicity of estrogen receptor agonists in a primary culture of rainbow trout (*Oncorhynchus mykiss*) hepatocytes. *Aquat Toxicol* 101: 186–195.

Putman, N.F., Abreu-Grobois, A., Iturbe-Dearkistade, I., Putman, E.M., Richards, P.M., Verley, P. 2015. Deepwater Horizon oil spill impacts on sea turtles could span the Atlantic. *Biol Lett* 11: 20150596.

Sharma, B.M., Bharat, G.K., Tayal, S., Nizzetto, L., Cupr, P., Larssen, T. 2014. Environment and human exposure to persistent organic pollutants (POPs) in India: A systematic review of recent and historical data. *Environ Int* 66: 48–64.

Smithsonian Institute Ocean Portal. 2014. Gulf oil spill. ocean.si.edu/gulf-oil-spill. Accessed November 23, 2016.

Stansley, W., Roscoe, D.E. 1999. Chlordane poisoning in birds in New Jersey, U.S.A. *Environ Toxicol Chem* 18: 2095–2099.

U.S. Environmental Protection Agency (EPA). 2014. Toxicological review of benzo[a]pyrene. (CASRN 50-32-8). Washington, DC: U.S. Environmental Protection Agency. https://yosemite.epa.gov/sab%5CSABPRODUCT.NSF/PeopleSearch/4DCFD0E5F45A8CAD 85257B65005B17C8/$File/Benzoapyrene_external+review+draft+IRIS+Toxicological +Review_HERO_9-25-14.pdf. Accessed December 12, 2016.

U.S. Environmental Protection Agency (EPA). 2016a. Priority pollutants. https://water.epa. gov/scitech/methods/cwa/pollutants.cfm. Accessed December 11, 2016.

U.S. Environmental Protection Agency (EPA). 2016b. Oil spills research. https://www.epa. gov/land-research/oil-spills-research. Accessed December 11, 2016.

U.S. Fish and Wildlife Service (FWS). 2011. Deepwater Horizon response consolidated fish and wildlife collection report. http://www.fws.gov/home/dhoilspill/pdfs/ ConsolidatedWildlifeTable042011.pdf. Accessed November 23, 2016.

Yordy, J.E., Wells, R.S., Balmer, B.C., Schwacke, L.H., Bowles, T.K., Kucklick, J.R. 2010. Life history as a course of variation for persistent organic pollutant (POP) patterns in a community of common bottlenose dolphins (*Tursiops truncatus*) resident to Sarsota Bay, FL. *Sci Total Environ* 408: 2163–2172.

Yordy, J.E., Rossman, S., Ostrom, P.H., Reiner, J.L., Bargnesi, K. et al. 2013. Levels of chlorinated, brominated, and perfluorinated contaminants in birds of prey spanning multiple trophic levels. *J Wildl Dis* 49: 347–354.

Zhang, Y., Tao, S. 2009. Global atmospheric emission inventory of polycyclic aromatic hydrocarbons (PAHs) for 2004. *Atmos Environ* 43: 812–819.

8 Contaminants of Increasing Concern

8.1 INTRODUCTION

You might want to call this chapter a 'catch all' because we are going to cover a small variety of contaminants that have not been discussed before but nevertheless are important. Now, it would be virtually impossible to cover all the different kinds of contaminants in a book that is intended to be introductory and therefore not necessarily comprehensive. For a more comprehensive text the avid student might want to check out one of the other textbooks that have been published including (warning: personal plug here) Sparling (2016). There are literally scores of other contaminants that we will not discuss, some with considerable research, some not so much. So, to wrap up this section dealing with specific types of contaminants, we will present some information on plastics, pharmaceuticals, and nanoparticles and end with acid deposition.

8.2 PLASTICS

According to the Environmental Protection Agency (EPA, 2015), 32 million tons of plastic waste were generated in 2012, representing 12.7% of total manufactured solid wastes. About 14 million tons of plastics were containers and packaging, 11 million tons were durable goods such as appliances, and the rest were nondurable goods, such as plates and cups. Of that total, only 9% was recovered for recycling despite the fact that 86% of the American public have access to either curbside or recycling center pickup (EPA, 2015). Of the plastics that are recycled, most include the types that go into cups and plates, followed by those convenient grocery store plastic bags and wraps.

One might ask, what, exactly, is a plastic? Well, that question requires a somewhat lengthy explanation. **Plastic** is a material consisting of any of a wide range of synthetic chemicals that can be molded into solid objects of diverse shapes. Most are polymers of high molecular mass, but they often contain other substances. For example, Figure 8.1 represents single polymer units of polyethylene and polystyrene, very common types of plastics. An item made of either plastic consists of millions of these units strung together. In fact, some have likened such plastic objects to very, very large molecules consisting of one immense strand of polymers. Plastics are mostly derived from petrochemicals, but many are at least partially derived from other materials. Malleability is a general property of most plastics that can irreversibly deform without breaking, but this occurs in a wide degree of variability. Take, for instance, plastic bottles, which can be easily deformed, and compare them to polyvinyl chlorine (PVC) plumbing pipes, which are far more rigid—both are plastics. There are

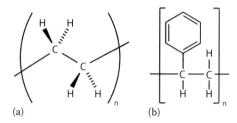

(a) (b)

FIGURE 8.1 Individual polymer units for polyethylene (a) and polystyrene (b). A given plastic object will have thousands to millions of these units strung together, depending on the size and density of the object.

various categories of plastics based on their composition. The numbers following the name are the recycling codes you find on plastic objects.

1. Polyethylene terephthalate (PETE) [1]—most often used for cooking oil bottles, soft drink bottles, and peanut butter jars; also, makes up polyesters or synthetic fibers such as Dacron. This type of plastic achieves the greatest amount of recycling. Note the word 'terephthalate'; we'll mention something about phthalates in a little while.
2. High-density polyethylene (HDPE) [2]—more rigid than low-density polyethylene, commonly used for milk jugs and detergent bottles. This is also a significant part of the recycling effort.
3. Polyvinyl chloride (PVC) [3]—used in plastic pipes, some water bottles, outdoor furniture, shrink-wrap, liquid detergent containers, and salad dressing containers.
4. Low-density polyethylene (LDPE) [4]—more flexible than high-density polyethylene, found in trash can liners, dry-cleaning bags, produce bags, and food storage containers. Plastics in this category are easily recycled.
5. Polypropylene (PP) [5]—used for drinking straws and bottle caps.
6. Polystyrene (PS) [6]—makes up packaging pellets like 'Styrofoam peanuts,' Styrofoam coolers, and some types of plastic drinking cups and plates.
7. Other [7]—Plastics listed in the OTHER category are those not listed in the first six categories. Certain food storage containers fall within this category.

Some plastics are the poster child for persistence. Scientists have argued about the permanence of plastics, with some suggesting that it may take hundreds of years for plastics to decompose, while others saying that some plastics may decompose more rapidly, and still others claiming that plastics never decompose, but just break into smaller and smaller pieces. Part of the controversy is that most plastics do not biodegrade; that is, microbes such as bacteria have no effect on them. However, plastics do break down. Given enough time, these pieces may become microscopic, but they are still in the environment. Also, there are a huge number of different types of plastics, some of which have biodegradable qualities. On land, polystyrene (think foam cups) may take up to 50 years or more to break down into unrecognizable small pieces. In oceans, however, recent studies (Saido et al., 2014) show that exposure to wind, sunlight, and water accelerates the decomposition of this plastic by a few years. Other plastics take even longer to decompose. Plastic water bottles may require 400 years to disappear, disposable diapers 450 years, and monofilament fishing lines 600 years (EPA, 2015). Often toxic chemicals such as polychlorinated biphenyls (PCBs), polybrominated diphenyl ethers (PCDEs), bisphenol A (BPA), and phthalates adhere to polyethylene and polypropylene and are released into the environment as the plastics degrade. These chemicals can cause problems unrelated to the plastics from which they came.

This enormous production of plastics and the low rate of recycling lead to a massive amount of plastics in our environment. Oceans, in particular, are the final repository for a lot of this pollution. Nearly a decade ago, an estimated 165 million tons (150 million tonnes) of plastics consisting of over 5 trillion pieces contaminated the ocean waters (Barnes et al., 2009); this amount undoubtedly has increased since then. "It is likely that nearly all of the plastic that has ever entered the environment still occurs as polymers and very little or any plastic fully degrades in the marine environment" (Andrady, 2009). About 20% of the plastics comes from oceanic sources such as dumping bilge water, discarded fishing gear, dumping cargo, and other sources of waste. Poorly maintained or designed landfills contribute a large number of plastic particles through leaching, resulting in discharge into rivers and streams and gradual movement downstream. Globally, the vast majority of plastics in the ocean, around 80%, comes from land (UNESCO, 2015).

There are two huge garbage patches largely composed of plastics but including some other debris in a subtropical convergence zone (Figure 8.2). Millions of tons of micro- (defined as particles <5 mm diameter), meso- (5–20 mm), and macroplastics (>200 mm) float on these patches. Andrew Turgeon (2014), a writer for National Geographic, stated that one expedition to the garbage patch collected 750,000 bits of plastic/km^2, which is equivalent to about 1.9 million bits/mile^2. According to the United Nations Environment Programme (UNESCO, 2015), in 2006, there were 46,000 pieces of floating plastics for every square mile of the ocean!

These plastics can cause problems for oceanic wildlife. A major concern for wildlife is meso- and macroplastics such as fishing lines and nets. UNESCO estimates that more than a million seabirds, 100,000 marine mammals, and an untold number of sea turtles are killed each year by getting caught in discarded or lost fishing nets (Figure 8.3). Other animals die because they ingest plastics that block their digestive systems. Plastic bags, the type you get when shopping, last for decades and look a lot like jellyfish to hungry sea turtles (Figure 8.4). When they are swallowed by the turtles they block gastrointestinal passages, resulting in mortality. Soda (or beer, if you prefer) ring can holders can also pose severe problems (Figure 8.5).

Two of the chemicals released by degrading plastics are known to be potent endocrine disruptors. One of these is **phthalates**. Recall that we said we would come back to this when we were discussing polyethylene terephthalate. Phthalates are used as solvents for other materials and in hundreds of products such as vinyl flooring, adhesives, detergents, lubricating oils, automotive plastics, plastic clothes (raincoats), and personal care products (soaps, shampoos, hair sprays, and nail polishes). They are

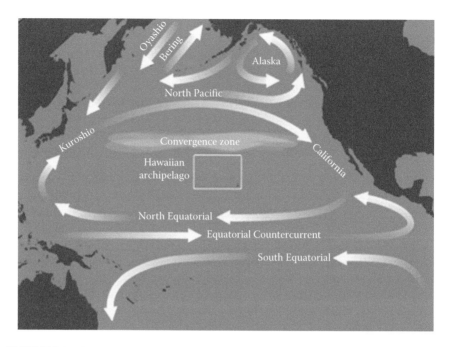

FIGURE 8.2 The Great Pacific Garbage Patch or convergence zone. (Courtesy of NOAA, Where are the Pacific garbage patches, 2013, http://marinedebris.noaa.gov/info/patch.html.)

FIGURE 8.3 An untold number of sea turtles along with seabirds and mammals die each year as they get tangled in fishing nets and lines. (Courtesy of West, P., NOAA scientists battle ocean ghostnets, NOAA magazine, 2005, http://www.noaanews.noaa.gov/stories2005/s2429.htm.)

FIGURE 8.4   It is hard for a turtle to distinguish between floating plastic bags and jellyfish. If they accidentally consume plastic, the bags can block the digestive system, resulting in death. (Courtesy of Wikimedia Commons, http://saynotoplastic.wikia.com/wiki/Plastic_bags_and_the_Environment?file=Plastics_jellyfish.jpg.)

FIGURE 8.5   Animals can get caught in plastic beverage holders and suffer grave consequences. This turtle survived to be the mascot of an organization that supports recycling (after the ring was removed, of course). (Courtesy of Messenger, S., Turtle cut free from 6-pack rings is unstoppable 20 years later, The Dodo, 2015, https://www.thedodo.com/turtle-six-pack-unstoppable-1166240209.html.)

widely used in PVC plastics as plastic packaging film and sheets, pipes, garden hoses, inflatable toys, blood-storage containers, medical tubing, and other products.

Humans usually are not exposed to phthalates in any meaningful concentration, so they do not cause concern among many federal regulatory agencies. However, the U.S. EPA cites studies that have associated phthalates with human health concerns including endocrine disruption (ED) and reduced male fertility; consequently, this agency is beginning to restrict the use of the chemicals (EPA, 2016a).

During laboratory studies on rodents, phthalates cause multi-organ damage through oxidative stress and DNA damage. They can interfere with many cell functions and antioxidant enzymes. Phthalates seem to target reproductive systems and cause disruption of spermatogenesis while they induce mitochondrial dysfunction in precursors to gametes (Asghari et al., 2015).

The other chemical that is released during the degradation of plastics is BPA. This is used to make some plastics and epoxy resins clear and strong. It is common in water bottles, sports equipment, food storage containers, CDs, DVDs, and lining of water pipes and as coatings on the inside of many food and beverage cans (e.g., the white inside lining seen in many cans).

BPA causes ED in laboratory animals and raises concern about its suitability in some consumer products and food containers. In 2012, the U.S. Food and Drug Administration banned its use in baby bottles and toddler cups; Canada and the European Union have also banned BPA in those uses.

In a review of BPA in wildlife, Crain et al. (2007) found extensive evidence that BPA induces feminization during the development of gonads in fishes, reptiles, and birds. Also, there is evidence that exposure of adult male fish can be detrimental to spermatogenesis and stimulates vitellogenin synthesis in fish. Vitellogenin is a serum protein involved in the production of egg yolk; its presence in males suggests feminization of the animals. In all cases, the amount of BPA necessary to cause such disruption in acute exposures exceeded environmentally realistic concentrations. Since chronic exposures to BPA cause effects at lower concentrations than acute exposures, the authors concluded that long-term exposure to BPA could exert reproductive effects at environmentally realistic concentrations. Further, they concluded that there was no concentration that could be considered safe to aquatic animals under chronic exposure.

8.3 PHARMACEUTICALS

Every time you go to the bathroom you risk polluting the environment with drugs. At any given time, 28% of all women aged 15–44, a total of more than 26 million people, use birth control pill or other hormonal treatments to prevent pregnancy (CDC, 2016). These medications contain female hormones, notably progestins and estrogens. Approximately 27 million people have Type II diabetes (CDC, 2014), and many of these use metformin or other medications to treat their disease. Almost everyone uses antibacterial soaps, hand cleansers, and other products that contain triclosan, triclocarban, or chloroxylenol. Pet care products contain many of the same ingredients that may be found in products for humans. These and hundreds of other drugs including antibiotics, antidepressants, anti-inflammatory analgesics, beta-blockers, and hormone replacement therapies are eliminated through urination or defecation or are rinsed off into the sink and enter sewage treatment facilities. However, sewage treatment plants are ill-equipped to handle these pharmaceuticals and end up disposing them into natural waterways.

The extent of contamination from these products and the possible effects they might cause are still not well known. Collectively, these pharmaceutical and personal care products are referred to as PPCPs. Several studies have documented scores of PPCPs in various sample areas. For example, Yu et al. (2006) found

16 of 18 PPCPs they analyzed in the Black River Wastewater Treatment Plant of Baltimore, MD. These chemicals spanned a range of therapeutic classes and personal care products in raw sewage. The highest concentrations in river water were found for ibuprofen (1.9 µg/L), naproxen (3.2 µg/L), and ketoprofen (1.2 µg/L), all of which are pain-reducing nonsteroidal anti-inflammatory drugs (NSAIDs). Ten of the 18 selected PPCPs were detected in the effluent discharge coming from the sewage treatment plant, indicating that these chemicals were not completely removed from the sewage.

Another source of pharmaceuticals to the environment are landfills. Masoner et al. (2014) identified 129 PPCP chemicals in untreated leachate samples from 19 landfills. The chemicals most frequently found during this study included BPA (from plastics), cotinine (a by-product of tobacco), N,N-diethyltoluamide (DEET, an insect repellent), lidocaine (local anesthetic), and camphor (mostly a skin treatment), all of which occurred in 90% or more of the landfill leachates. Concentrations of these chemicals were in the ng/L to µg/L range.

Many of these pharmaceuticals can concentrate in tissues. Of 114 PPCPs and artificial sweeteners, 83 had biological concentration factors (BCFs) greater than 1 and 15 had BCFs greater than 100 (Lazarus et al., 2015).

Perhaps the worst case of PPCP-induced mortality in wildlife to date befell vultures in India (Oaks et al., 2004). Severe population declines by 95% or more started with the Oriental white-backed vulture (*Gyps bengalensis*) and spread to catastrophic declines of two other species, *G. indicus* and *G. tenuirostris*. These once common vultures were declared critically endangered by the International Union for Conservation of Nature (IUCN). During postmortem examinations, Oaks et al. (2004) found that of 259 adult and subadult *G. bengalensis*, 219 (85%) had visceral gout on the surface of internal organs. Visceral gout in birds is most commonly the result of renal failure leading to the deposition of uric acid on and within the internal organs. Necropsies of recently dead vultures with visceral gout revealed acute renal failure due to a toxic cause along with the gout. After extensive elimination of possible alternative causes, scientists concluded that ingested diclofenac, a NSAID, might be responsible for the renal disease in the birds. Diclofenac was extensively used as a veterinary treatment for ailing cattle. Diclofenac was subsequently confirmed as the cause through experimental dosing of a few birds; three of four developed visceral gout and kidney failure. The authors even fed captive vultures meat from cattle and goats that had been dosed with diclofenac and several of these birds developed gout and kidney failure in a dose–response manner. Subsequently, diclofenac was banned in the Indian subcontinent for veterinary use and vulture populations seem to be making a slow comeback.

8.4   NANOPARTICLES

By definition, nanoparticles are very small, 1–100 nanometers (nm) in size, small, but larger than most molecules. Their small size conveys characteristics that differ from their larger or bulk versions. One very important difference is that nanoparticles have a very large surface-to-volume ratio. For example, a microparticle of carbon atoms, which is about 1000 times larger than a nanoparticle, may have a mass of 0.3 µg and a surface area of 0.01 mm^2. But the same mass of carbon in nanoparticle form has a surface area of 11.3 mm^2 and consists of 1 billion nanoparticles. With this large surface area, chemical reactions can be 1000 times faster than with microparticles (Buzea et al., 2007). In addition to fast chemical reactions, nanoparticles often possess unexpected optical and magnetic properties because they can produce quantum effects. For example, we are used to the characteristic yellowish color and high sheen of gold, but gold nanoparticles appear deep red to black in solution. Gold nanoparticles also melt around 300°C, but bulk gold does not melt until 1064°C (Buffat and Borel, 1976). Also, nanoparticles absorb solar radiation more efficiently than bulk particles, which can be effective as sun blockers in building materials. In fact, to understand the behavior of nanoparticles, one must apply quantum mechanics (Buzea et al., 2007), and that is above my level of understanding.

You might think that nanotechnology is something new, but nanoparticles have been used since ancient times and they are environmentally common. In the ninth century BCE, potters in Mesopotamia used glazes containing nanoparticle materials to produce a glittering effect. Similar glazes were used during the Middle Ages (Khan, 2012). Nanoparticles are just about everywhere—in dust, volcanic emissions, ash, and many other natural processes. Even viruses are nanoparticles. Because of the prevalence of nanoparticles, organisms, including humans, have developed multiple defense mechanisms including a thick skin mucous secretion, and antibodies that attack foreign proteins. Once in the body, however, particles in the nano- category can freely enter cells.

In this section, however, we are mainly concerned with the much more recent explosion of nanoparticle production in many industrial applications. Many of the nanoparticles now used in manufacturing, medicine, and other areas are composed of the same materials used in bulk particles, but in this case, size does matter. We are only experiencing the tip of a potential iceberg in the number of uses of nanoparticles. Nanoparticles are being or will soon be used in the food preservation and packaging industry to provide monitors of bacteria and food quality, to preserve freshness through nanocoatings that provide antibacterial agents, and to directly produce foods. Nanoparticles are also being used to provide ultrasmooth surfaces in ceramics and in ultrathin polymers on sunglasses, to produce nanofibers that enhance stain resistance and wrinkle-free clothing, and to produce very effective sunburn lotions. In sports, baseball bats are being made from carbon nanotubes to increase their strength while reducing their weight; sport towels, yoga mats, and exercise mats that contain antimicrobial technology are on the market.

In medicine, nanoparticles are used in imagery dyes, fluorescent chips, and quantum dots that emit energy that can identify real or potential illnesses. Nanoliposomes are ultrasmall containers that can deliver medicines to specific areas of the body (Figure 8.6). These liposomes and other nanoparticles are being used to treat cancers, HIV/AIDS, neurodegenerative disorders, ocular ailments, and respiratory ailments (Murthy, 2007).

Nanotubules are very strong and very lightweight containers made of carbon and have broad military applications including making uniforms that are resistant to bullets yet lightweight. They also bend light waves to make the so-called invisible camouflage, which reflects background colors and patterns and is now in early stages of testing. The list of possibilities is awe inspiring.

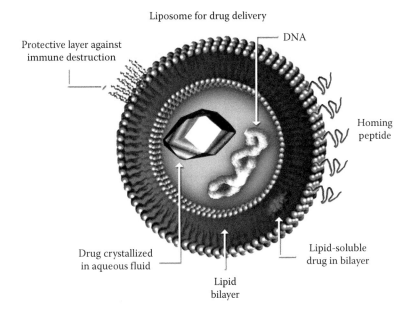

FIGURE 8.6 Cross section of a hypothetical liposome. Liposomes are composed of nanoparticles and can carry DNA or medicine to specific tissues. Homing peptides are specific for the target tissues, protective layers reduce detection by antibodies, and the lipid layer facilitates penetration into a cell. (Courtesy of Torchilin, V., *Adv. Drug Deliv. Rev.*, 58, 1532, 2006.)

Clearly, nanos are the next-generation materials. But as the use of nanoparticles increases, should we be asking "Is there need for concern about the widespread use of particles that behave in ways that we do not fully understand as yet?" For several reasons the answer is "Yes, we should be concerned." Currently, there is much to learn about the effects, fate, and transport of nanoparticles in the environment. Certainly, there are studies being conducted to further develop our understanding of these matters, but the scope of applications and the magnitude of future production of the vast array of materials demand that we gain this knowledge before environmental damage occurs.

Buzea et al. (2007) summarized some factors to consider with regard to the toxicity of nanoparticles. Nanoparticles can enter an organism through the lungs (or gills), skin, and digestive tracts. Due to their small size, nanoparticles can easily move from these entry points into the blood and lymph systems and then to tissues throughout the body. Take, for example, the average diameter of a hepatocyte or liver cell, which is 20–30 μm, roughly 1000 times the size of a nanoparticle! That's like comparing an African elephant to a Chihuahua! Because they are so small and can be constructed to be nonpolar (without an external electrical charge), nanoparticles can work their way through pores in cell membranes or brain barriers and gain access more readily than even dissolved polar ions (Murthy, 2007). We know that some nanoparticles can produce oxidative stress or physical injury to cells. There is potential harm to organelles, nuclei, and DNA.

The most extreme health hazards are due to inhalation of nanoparticles (in general, not necessarily only manufactured particles), which has been linked to asthma, bronchitis, emphysema, lung cancer, and Parkinson's disease (Buzea et al., 2007). Ingested nanoparticles are associated with Crohn's disease and colon cancer. Nanoparticles present in the circulatory systems are related to occurrences of arteriosclerosis, blood clots, arrhythmia, heart diseases, and ultimately cardiac arrest. Nanoparticles have also been linked to immunodeficiency. These diseases have been mostly studied in relation to human health, and maladies possibly incurred by wild animals are less well known. This clearly is an area of research offering substantial promise.

There are a few studies that have examined the toxicity of nanoparticles in wildlife. Gaiser et al. (2012) compared the effects of silver at nominal diameters of 35 nm, 0.6 μm, and 1.6 μm on a few species of wildlife. The smallest particles were in the nano range, but the large two were in the micro range. They found that the nano range particles were more toxic than either of the micro range particles. The same study showed that during a 21-day chronic test with carp (*Cyprinus carpio*), nano silver particles attained higher concentrations in the liver, intestines, and gall bladder than did microparticles. The authors also demonstrated that in a test using cellular cultures of rainbow trout hepatocytes, nano silver was six times more toxic than micro silver.

8.5   ACID DEPOSITION

Acid deposition is the precipitation, whether in dry particulates or in rain or snow, of chemicals that cause acidic conditions. The most common of these chemicals include sulfur dioxide (SO_4^{2-}), nitrite (NO_2^-), and nitrate (NO_3^{2-}), for shorthand nitrites and nitrates are often collectively called nitrogen oxides and symbolized by NO_x. The root cause for acid deposition started with the Industrial Revolution in the mid-1700s and developed highly by the mid-1800s. Through that time and continuing until today, many contaminants have bellowed from the smokestacks of every coal-fired utility plant and many industries in North America and northern Europe. The most common include carbon dioxide, mercury, and lead, in addition to the ions already mentioned. Fortunately, much of the contaminant emissions have been greatly reduced due to mandatory smokestack scrubbers and other technologies. Still, roughly 2/3 of all SO_2 and 1/4 of NO_x in the atmosphere come from fossil fuel combustion for electricity, and coal-fired plants are the main contributors of these chemicals (EPA, 2016b).

In the atmosphere, sulfur dioxide combines with oxygen to form sulfur trioxide, which then is deposited into bodies of water such as lakes, and combines with water to produce sulfuric acid (Figure 8.7). Similarly, nitrogen oxide combines with oxygen in the atmosphere to form nitric acid. The process of acidification is sometimes incorrectly called acid rain. However, acid-causing ions can be mixed with any form of moisture such as rain, sleet, fog, or mist and be called **wet deposition**. These chemicals can also adhere to dust particles in the atmosphere and come down as **dry deposition**.

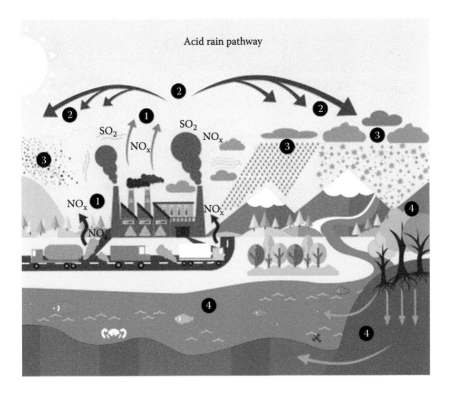

FIGURE 8.7 Acid deposition. (1) Emissions of SO_2 and NO_x are released into the air where (2) the pollutants are transformed into acid particles that may be transported long distances. (3) These acid particles then fall to the earth as wet and dry deposition (dust, rain, snow, etc.) and (4) may cause harmful effects on soil, forests, streams, and lakes. (Courtesy of U.S. Environmental Protection Agency (EPA), Effects of acid rain—Surface waters and aquatic organisms, 2016c, https://www.epa.gov/acidrain/what-acid-rain, accessed November 24, 2016.)

Starting in the 1960s and continuing through the 1980s, existing smokestacks on electrical generating plants were replaced with much taller ones to move emissions to higher elevations and supposedly remove them from the vicinity of humans. However, these stacks simply placed the emissions higher into the atmosphere where they could travel longer distances before depositing their acid-causing ions. Automobile exhausts and other sources of fossil fuel combustion added to the air-blown pollution. Cities in the Midwest were important sources of this pollution, but the effects were not experienced until hundreds of miles later, usually as the contaminant-laden air cooled over northeast United States and southeast Canada. Soils in the Northeast are largely shallow and granitic and have poor buffering capacity or **acid-neutralizing capacity (ANC)**. Wet deposition, often having a pH of 3, the same as vinegar, and dry deposition fall onto soils where the acid-forming molecules can flush into lakes, ponds, and streams, reducing their pH.

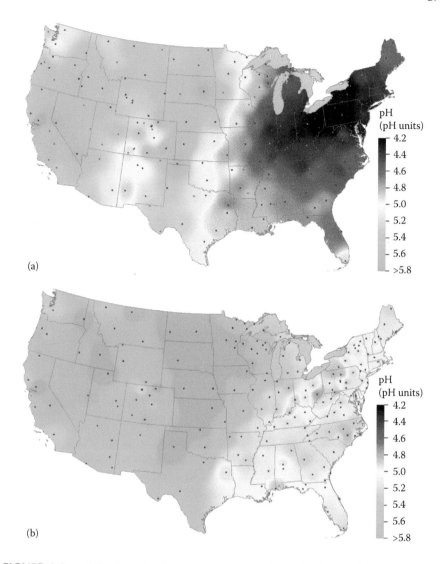

FIGURE 8.8 Acidity in water bodies measured as pH in 1989–1991 (a) compared to 2013 (b). (Courtesy of EPA CASTNET, Clean Air Status and Trends Network (CASTNET), 2017, http://epa.gov/castnet/javaweb/precipchem.html.)

The Midwest was less affected by its own acid deposition because much of the soil there has a limestone or calcium carbonate base and the calcium carbonate buffers sulfuric acids. In addition, Midwestern soils have natural organic acids such as tannic or humic acids that also help buffer soils.

The regulations that were put in place some 40 years ago have been effective. The total acidity in the atmosphere (Figure 8.8) and deposition of nitrogen oxides and sulfate have been substantially reduced in the United States compared to preregulatory times. At the same time, the pH of water bodies has increased, sometimes with the help of adding tons of limestone to streams and lakes.

Acid deposition has had significant harmful effects on plants, animals, and structures. Technically speaking, solutions are acidic if their pH drops below 7.0 and alkaline if the pH is above that mark. In practice, however, the pH of water is considered **circumneutral** and safe for most organisms at a range of 6.5–8, with some experts going as low as pH 6.0. Organisms are often at risk when pH levels drop below 6.0. For example, both aquatic and terrestrial snail shells, which are mostly calcium carbonate, begin to dissolve in acidic conditions. Snails cannot survive without their shells and begin to disappear when soil pH drops to 6 or below. Reduced snail abundance can, in turn, affect their predators. For example, Graveland et al. (1994) documented a decline in great tits (*Parus major*) and other passerines in the forests of the Netherlands. The declines were associated with thin eggshells. This time the eggshell thinning was not due to DDT derivatives but deficient calcium in the soil. Low pH also dissolves calcium and leaches it from soils. Snails, which are an important prey for great tits, require high soil calcium for their shells, and birds require it for their eggshells. With the leaching of calcium from the soil and diminished snail populations, the birds suffered reduced reproductive success. Further proof for a link between snails and bird populations was demonstrated when calcium was augmented in soils and snail populations, territories of ovenbirds (*Seiurus aurocapilla*) increased in density, clutch sizes increased, and more nests were built (Pabian and Brittingham, 2011).

Fish, clams, mussels, and aquatic snails and early signs of ecosystem stress are observed at pH below 6.0, in some cases 6.5. At pH 5.5, largemouth bass (*Micropterus salmoides*) begin to suffer and rainbow trout die due to problems with ion regulation and enzyme functioning (EPA, 2016c). Embryonic amphibians may be unable to hatch at pH around 4.5–5.0 or below because the membrane that surrounds the egg hardens and prevents the embryo from hatching; continued growth causes the body of the embryo to form a v-shape as it fills the space within the membrane until death occurs (Clark and Hall, 1985).

Oceans also experience acidification but through carbon dioxide mixing with water and producing carbonic acid (NOAA, n.d.). The average pH of ocean waters since roughly 1850 has dropped from 8.2 to 8.1. While this may seem small, the pH scale is logarithmic, which means that there has been a nearly 30% increase in hydrogen ion concentration (NOAA, 2016). If the output of 'greenhouse gases' continues, we may see several ramifications of increased ocean acidity such as making calcium less available in oceans. Organisms that rely on calcium for their shells and reef building corals may suffer (NOAA, 2016).

Plants may also suffer. For instance, coniferous forests in the Northeast United States, eastern Canada, and northern Europe have been severely harmed by acid deposition (Figure 8.9). Acidification can leach important minerals such as calcium and magnesium from the soil, making them unavailable to plants. Loss of nutrients weakens trees and other plants, making them more susceptible to other factors such as diseases. In addition, acidification changes elemental aluminum, the most abundant metal in soils, to more toxic ionic forms.

FIGURE 8.9   Forest dieback in Germany due to acid deposition. (Courtesy of Lovecz, https://commons.wikimedia.org/w/index.php?curid=938073.)

8.6   CHAPTER SUMMARY

1. Each year Americans use tons of plastics, of which only 9% are recycled. The rest of the plastic goes into landfills, incinerated, or tossed away as a form of pollution.

2. Plastics are among the most persistent of all contaminants. Many exist in the environment for hundreds of years or longer. They may break down into small pieces called microplastics but still exist essentially unchanged by chemical reactions. Few microorganisms have any ability to biodegrade plastics.

3. During their presence in the environment, plastics may give off various chemicals including PCBs, and pesticides that are present in plastic containers plus they emit the known endocrine disruptors, BPA and phthalates.

4. A lot of the plastics coming from both marine and terrestrial sources have accumulated in two Great Garbage Patches in the Pacific Ocean.

5. Bulk plastics can kill wildlife by trapping them in nets, causing compaction in the digestive system and other ways. Ingested pieces of plastics can also cause compaction, but since plastic is not digestible, the plastic material itself is not exceptionally toxic.

6. Hundreds of PPCPs end up in natural waterways through a complex process involving flushing them down toilets or rinsing them in sinks. They then enter sewage treatment plants that are not equipped to treat these products and empty them through effluent into waterways.

7. PPCP can cause many problems with wild organisms, including endocrine disruption, genotoxicity, and various other disease conditions.

8. Nanoparticles are very small bits of carbon, metals, and other materials that are finding rapidly increasing applications in many different industries. The variety of uses is just beginning to be realized, but with use comes a great potential for these particles to enter the environment.

9. Currently, the level of knowledge about the possible effects of nanoparticles in the environment is not very great. A few studies have shown that they can harm wildlife, but a lot more work needs to be done.

10. The source of acid deposition began over 250 years ago with the start of the Industrial Revolution. Coal-fired electrical plants and various other industries produce gases through smokestack emissions that produce acidic conditions when mixed with water.

11. The areas most severely hit by acid deposition include northeastern United States, southeastern Canada, and northern Europe.

12. Acidic conditions caused by this deposition cause many problems on land and in water. Generally, pH conditions below 6.5 or 6.0 are causes for concern. Organisms that use calcium for their shells are at jeopardy because calcium dissolves in acidic conditions. Other aquatic species suffer ion imbalances or other maladies. Forests have also suffered in Europe due to acidification and the resulting dissolved aluminum concentrations.

8.7   SELF-TEST

1. Which of these contaminants is expected to survive the longest in the environment?
   a.   Plastics
   b.   PCBs
   c.   DDT
   d.   PBDEs
2. Short answer. In your own words, describe what differentiates plastics from other common environmental contaminants.
3. What is the principal concern for the toxicity of BPA and phthalates?
   a.   Cardiac disorders
   b.   Genotoxicity
   c.   Endocrine disruption
   d.   Malformations
4. True or False. When an animal is exposed to a pharmaceutical under natural conditions, it is most likely exposed to only one contaminant at a time.
5. Vultures in India have suffered catastrophic losses due to what type of veterinary medicine?
   a.   An antibiotic
   b.   A synthetic estrogen
   c.   A painkiller
   d.   None of the above

6. What is the largest diameter a particle can have and still be considered a nanoparticle?
   a. 100 nm
   b. 10 nm
   c. 10 µm
   d. 100 mm
7. Short answer. Describe what a liposome is and what it can do.
8. True or False. Even smaller nanoparticles are too large to fit through pores in cell membranes.
9. Short answer. What ions from smokestack emissions are most responsible for acid deposition?
10. A pH of __ is a cause for environmental concern.
    a. 8.0
    b. 7.0
    c. 6.8
    d. 6.0

REFERENCES

Andrady, A. 2009. Fate of plastics debris in the marine environment. In: Arthur, C., Baker, J., Bamford, H. (eds.), *Proceedings of the International Research Workshop on the Occurrence, Effects, and Fate of Microplastic Marine Debris*, University of Washington Tacoma, Tacoma, WA, September 9–11, 2008. NOAA Technical Memorandum NOS-OR&R-30.

Asghari, M.H., Saeidnia, S., Abdollah, M. 2015. A review on the biochemical and molecular mechanisms of phthalate-induced toxicity in various organs with a focus on the reproductive system. *Int J Pharmacol* 11: 95–105.

Barnes, D.K.A., Galgani, F., Thompson, R.C., Barlaz, M. 2009. Accumulation and fragmentation of plastic debris in global environments. *Philos Trans R Soc B Biol Sci* 364: 1985–1998.

Buffat, P.H., Borel, J.P. 1976. Size effect on the melting temperature of gold particles. *Phys Rev A* 13: 2287–2298.

Buzea, C., Pacheco, I.I., Robbie, K. 2007. Nanomaterials and nanoparticles: Sources and toxicity. *Biointerphases* 2: MR17–MR71.

Center for Disease Control and Prevention (CDC). 2014. Diabetes latest. http://www.cdc.gov/features/diabetesfactsheet/. Accessed November 24, 2016.

Center for Disease Control and Prevention (CDC). 2016. Contraceptive use. http://www.cdc.gov/nchs/fastats/contraceptive.htm. Accessed November 24, 2016.

Clark, K.L., Hall, R.J. 1985. Effects of elevated hydrogen-ion and aluminum concentrations on the survival of amphibian embryos and larvae. *Can J Zool* 63: 116–123.

Crain, D.A., Eriksen, M., Iguchi, T., Jobling, S., Laufer, H. et al. 2007. An ecological assessment of bisphenol-A: Evidence from comparative biology. *Reprod Toxicol* 24: 225–239.

Gaiser, B.K., Fernandes, T.F., Jepson, M.A., Lead, J.R., Tyler, C.R. et al. 2012. Interspecies comparisons on the uptake and toxicity of silver and cerium dioxide nanoparticles. *Environ Toxicol Chem* 31: 144–154.

Graveland, J., Vanderwal, R., Vanbulen, J.H., Vannoordwijk, A.J. 1994. Poor reproduction in forest passerines from decline of snail abundance on acidified soils. *Nature* 369: 446–448.

Khan, F.A. 2012. *Biotechnology Fundamentals*. Boca Raton, FL: CRC Press.

Lazarus, R.S., Rattner, B.A., Brooks, B.W., Du, B., McGowan, P.C. et al. 2015. Exposure and food web transfer of pharmaceuticals in ospreys (*Pandion haliaetus*): Predictive model and empirical data. *Integr Environ Assess Manage* 11: 118–129.

Masoner, J.R., Kolpin, D.W., Furlong, E.T., Cozzarelli, I.M., Gray, J.L., Schwab, E.A. 2014. Contaminants of emerging concern in fresh leachate from landfills in the conterminous United States. *Environ Sci Process Impacts* 16: 2335–2354.

Messenger, S. 2015. Turtle cut free from 6-pack rings is unstoppable 20 years later. The Dodo. https://www.thedodo.com/turtle-six-pack-unstoppable-1166240209.html.

Murthy, S.K. 2007. Nanoparticles in modern medicine: State of the art and future challenges. *Int J Nanomed* 2: 129–141.

National Oceanic and Atmospheric Administration (NOAA). 2016. PMEL Carbon Program. What is ocean acidification? http://www.pmel.noaa.gov/co2/story/What+is+Ocean+Acidification%3F. Accessed November 24, 2016.

NOAA. 2013. Where are the Pacific garbage patches. http://marinedebris.noaa.gov/info/patch.html.

Oaks, J.L., Gilbert, M., Virani, M.Z., Watson, R.T., Meteyer, C.U. et al. 2004. Diclofenac residues as the cause of vulture population decline in Pakistan. *Nature* 427: 630–633.

Pabian, S.E., Brittingham, M.C. 2011. Soil calcium availability limits forest songbird productivity and density. *Auk* 128: 441–447.

Saido, K., Koizumi, K., Sato, H., Ogawa, N., Kwon, B.G. et al. 2014. New analytical method for the determination of styrene oligomers formed from polystyrene decomposition and its application at the coastlines of the North-West Pacific Ocean. *Sci Total Environ* 473: 490–495.

Sparling, D.W. 2016. *Ecotoxicology Essentials: Environmental Contaminants and Their Biological Effects on Animals and Plants*. Cambridge, MA: Academic Press.

Torchilin, V. 2006. Multifunctional nanocarriers. *Adv Drug Deliv Rev* 58: 1532–1555.

Turgeon, A. 2014. Great Pacific garbage patch. http://nationalgeographic.org/encyclopedia/great-pacific-garbage-patch/. Accessed November 24, 2016.

United Nations Educational, Scientific and Cultural Organization (UNESCO). 2015. Facts and figures on marine pollution. http://www.unesco.org/new/en/natural-sciences/ioc-oceans/focus-areas/rio-20-ocean/blueprint-for-the-future-we-want/marine-pollution/facts-and-figures-on-marine-pollution/. Accessed November 24, 2016.

U.S. Environmental Protection Agency (EPA). 2015. Plastics. https://www3.epa.gov/epa-waste/conserve/tools/warm/pdfs/Plastics.pdf.

U.S. Environmental Protection Agency (EPA). 2016a. Phthalates. https://www.epa.gov/assessing-and-managing-chemicals-under-tsca/phthalates. Accessed November 24, 2016.

U.S. Environmental Protection Agency (EPA). 2016b. What is acid rain? http://www.epa.gov/acidrain/what/. Accessed December 14, 2016.

U.S. Environmental Protection Agency (EPA). 2016c. Effects of acid rain—Surface waters and aquatic organisms. https://www.epa.gov/acidrain/what-acid-rain. Accessed November 24, 2016.

West, P. 2005. NOAA scientists battle ocean ghostnets, NOAA magazine. http://www.noaanews.noaa.gov/stories2005/s2429.htm.

Yu, J.T., Bouwer, E.J., Coelha, M. 2006. Occurrence and biodegradability studies of selected pharmaceuticals and personal care products in sewage effluent. *Agric Water Manag* 86: 72–80.

Section III

**Higher Level Effects,
Analysis of Risk, and
Regulation of Chemicals**

9 Studying the Effects of Contaminants on Populations

9.1 INTRODUCTION

Before we get into specific ways that contaminants may affect populations, both real and theoretical, we need to understand that it is often very difficult to determine if a contaminant or group of contaminants actually causes a decline or extirpation of a population from an area. Sure, there have been very clear examples of contaminants leading to population declines. The reduction in the numbers of fish-eating birds coinciding with the use of organochlorine pesticides; the confirmation that dichlorodiphenyldichloroethylene (DDE) caused eggshell thinning seen in many nesting populations of brown pelicans (*Pelecanus occidentalis*), bald eagles (*Haliaeetus leucocephalus*), and osprey (*Pandion haliaetus*) in North America and elsewhere, and the restoration of these birds after the banning of DDT prove that members of the DDT family caused these reductions. Add to that the documentation that diclofenac decimated populations of *Gyps* vultures in India, the pesticide poisoning of birds in New Jersey (Stansley and Roscoe, 1999), and hundreds of other instances that have occurred, and the concept would seem to be irrefutable. Yet, there are still naysayers who claim that other factors led to the near demise of these populations. And that is the crux of the issue. Wildlife populations face all sorts of problems. They face competition from other members of their species for food, nesting areas, and other resources. They are exposed to diseases and parasites. Unless they are at the top of their food chains (and even then their young may be vulnerable), they run the risk of being eaten. Then there are the more unpredictable factors like weather and fire. In short, contaminants are usually only one of the many risks wildlife encounter. The trick is determining how much of an effect contaminants have and whether their effect is sufficient to topple the population.

In trying to put this into some sort of context that can aid our understanding and research, scientists (e.g., Burnham and Anderson, 1984) have developed the concepts of **additive and compensatory mortality**. A lethal factor such as predation, disease, or contaminants is a cause of additive mortality if and only if the deaths it causes would not have occurred by other factors in the absence of the lethal factor. In other words, the factor under consideration caused unique deaths that disease, predation, etc., would not have caused at the time (see Figure 9.1). Compensatory mortality occurs when the animals (or plants, I suppose) killed by contaminants would have died around that time anyway because they were eaten, diseased, frozen, or for some other reason. As far as proving that a given factor really alters population dynamics, additive mortality is the gold standard. The trick is trying to prove that contaminants (or whatever) are additive. Finding proof of additivity in the case of contaminants is

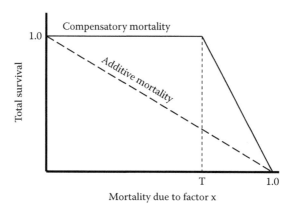

FIGURE 9.1   Conceptual model showing the difference between compensatory and additive mortality. With compensatory mortality, no real population effect occurs until some, probably very high, level. With additive mortality, actual population effects may be seen even at the lowest effects level.

further complicated in that environmentally realistic concentrations of most contaminants are well below that needed to kill animals acutely. Rather, they are subtle in reducing reproductive output or making the animals sick or weakened. If a rabbit that has been exposed to carbamate is having spasms and convulsions and a fox comes by and eats the rabbit, is the lethal factor the pesticide from which the rabbit may or may not have recovered or the fox who presumably would not have found such an easy meal without the pesticide?

In a real-life example, I have done a fair amount of work on declining frog populations in California (e.g., Sparling et al., 2001). To keep a long story short, there has been a long history of multiple contaminants being air blown from the Central Valley of California, one of the most productive agricultural centers in the country, to the foothills and middle elevations of the Sierra Mountains. We have even shown that tadpoles in more contaminated sites have a higher mortality rate than those in less contaminated areas, that their bodies carry up to 10 different contaminants, and that the enzyme acetylcholinesterase, which is deactivated by carbamates and organophosphate pesticides, is lower in these highly contaminated sites (Sparling et al., 2015). Nevertheless, critics, with some justification, have stated that we have not shown a true contaminant-caused population effect because frogs have a high fecundity and we do not know if the tadpoles would have died anyway. Complicating the picture even more, there is a fungus, *Batrachochytrium dendrobatidis*, that is known to cause overt, subacute death among adult frogs in the same areas. Seeing a pond littered with dead frogs is pretty convincing that the fungus is potent. We would somehow have to factor out the other causes of mortality, including diseases to prove that the populations were being reduced by the toxins in their water.

Having said all that, let us now take a look at some characteristics of populations and discuss whether and how contaminants might have an effect on them. Such a speculation is what field biologists might do to give themselves a reference point for diagnosing mortality due to contaminants.

9.2 HOW MIGHT CONTAMINANTS INFLUENCE THE CHARACTERISTICS OF POPULATIONS?

In this section we are going to look at some general characteristics of populations. After we briefly define them, we will discuss some ways in which contaminants may influence these characteristics. This will involve some conjecture but also sound reasoning. Following this, we will include some specific examples of studies that have documented population effects. In some ways, populations are like individuals, but instead of describing single events such as birth and death, when we speak of populations, we typically look at rates or ratios such as natality and mortality rates. An individual is of a certain sex, populations have sex ratios; individuals are a certain age, populations have an age structure; I'm sure you get the idea.

9.2.1 DENSITY DEPENDENCE AND INDEPENDENCE

These terms refer to how environmental factors affect populations. In many ways we can consider environmental contaminants to be another stress factor along with predation, disease, and the like. Density dependent factors are ones whose strength varies with the density of the population. Density independent factors do not (Figure 9.2). Both types of factors can negatively impact a population, even severely, but only density dependent factors can control population levels at some sort of steady state.

Usually we are talking about factors that have a negative effect on populations, and in that case density dependent factors affect a greater proportion of the population at high densities than at low densities. Predation can be a density dependent factor if predators increase the proportion of a population that they kill at higher densities of prey. Increases in predation can be either numerical when a greater number of predators are attracted to a high prey density or functional when existing predators key in on a prey species and take more of them without necessarily increasing in numbers.

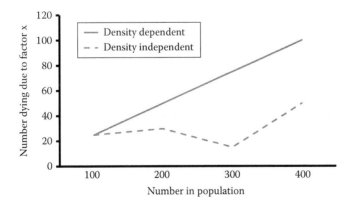

FIGURE 9.2 Density dependent and independent factors. In this diagram, the blue line indicates a density dependent negative factor. As the number in the population increases, the strength, measured by the number dying, also increases. The orange curve illustrates a density independent factor whose strength is independent of the population size.

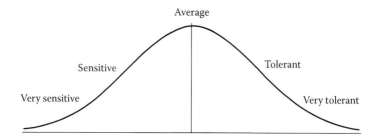

FIGURE 9.3   Generalized curve showing the response of different members of a population to a given contaminant. Although not shown, the concentration or dose of the contaminant increases along the horizontal axis. Very sensitive members of the population are affected at very low concentrations; the average members are affected when concentrations increase, but even higher concentrations are required to affect the most tolerant individuals in the population.

Keep in mind that we are talking about proportions of populations, not absolute numbers. For example, if predators (or any other mortality factor) take 50 individuals of a species when that species has 100 members (50%) and 60 individuals when the population has 200 members (30%), we see an increase in the absolute numbers, but the mortality is not in tune with density. To see a density dependent response, the predators would have to take at least 100 prey when its numbers equal 200. This is a simplistic explanation of some very important population characteristics. If you have not yet had a course in ecology, I would highly recommend that you do.

Depending on circumstances, contaminants can be either density dependent or independent, although in most cases we might expect them to be density independent. It may seem obvious but the proportion of a population that succumbs to a contaminant in large part depends on the concentration and toxicity of the contaminant. In any population, there will be some individuals that are very sensitive to the chemical(s), others that are very tolerant, and most somewhere in between (Figure 9.3). Thus, the greater the concentration of a contaminant with a given toxicity, the greater the proportion of the population that will be affected. This, arguably, is a form of density dependence common to all toxic contaminants. Given a single concentration, however, contaminants are likely to affect those individuals that are sensitive at that level and below but preserve the rest.

In addition to this, a given concentration of a pollutant may be density dependent, say, if there are safe spots in the environment where the concentration is lower. If the number of safe spots is limited at low populations, a large proportion of the population may be able to find them, but at higher populations, the safe spots may be occupied and those individuals not able to find security are exposed to lethal concentrations of the contaminant.

9.2.2   ABUNDANCE

It should go without saying that an individual is a single organism but populations are groups of organisms. In fact, a good working definition of a population is a group of interbreeding organisms of the same species occupying the same habitat or area. Abundance is the number of organisms—plant or animal—in that population, but unconstrained estimates of abundance are rather meaningless. A hundred animals could be localized in a $10\,km^2$ or a $1000\,km^2$, and it would make a very big difference to the organisms in the population and probably to the habitat as well. Therefore, ecologists usually refer to the density of a population, which is the number of organisms per unit area, such as 10 rabbits per acre ($=0.004\,km^2$).

Abundance and density can change through time. Populations can grow and diminish. Population growth is due to the interaction of four factors: (1) natality, (2) mortality, (3) emigration, and (4) immigration. **Natality** is birth and it is usually expressed as a rate, such as the number of young born per adult in the population or per the number of breeding females in a population. **Mortality** is death and is also expressed as a rate—either as a measure of time like the number of lemmings dying per year or as an aspect of the population as in 99% of amphibian tadpoles dying before they become adult frogs. **Emigration** is the number of animals (mature plants generally cannot emigrate or immigrate—although sometimes their seeds or spores may travel long distances) leaving a population and **immigration** is the number coming into a population from outside sources.

In short, population growth can be positive or negative and is modeled by the following equation:

$$\text{Growth} = \text{Natality rate} - \text{Mortality rate} + \text{Immigration rate} - \text{Emigration rate} \quad (9.1)$$

Contaminants can influence all of these factors of abundance. Mortality is the most obvious factor. We know that given a sufficiently high dose, most contaminants can be lethal, at least in the laboratory, but here we are concerned with natural populations. The most obvious example of this is when the mortality is intentional, such as the use of herbicides or insecticides. Mortality can also be unintentional. We have already shown that acid deposition has killed conifer forests in Europe and Canada (Chapter 8), the NSAID diclofenac in vultures in India (Chapter 8), and the effects of DDT compounds on fish-eating birds (Chapter 7). There are many other studies that can be cited, but sometimes we have to revert to using seminatural conditions such as mesocosms to at least partially control the other factors of mortality to determine what effect contaminants have.

One such seminatural experiment was conducted by de Solla et al. (2014) using eggs of snapping turtles (*Chelydra serpentina*). They wanted to see what effect pesticides used on potatoes in Canada would have on turtle nests. They placed the eggs at typical depths in soil within an outdoor enclosure and exposed them to one of five pesticides. Four of the chemicals, all insecticides, had no effect on the embryos and hatching rates were typical of wild nests. Metam sodium, a soil fumigant, however, killed all embryos even at concentration 1/10 of the prescribed application rate. This would be an example of contaminants affecting natality.

Sheffield and Lochmiller (2001) examined the effects of a pesticide, diazinon, on population dynamics of small mammals in a grassland setting. Again, they had to resort to seminatural conditions by blocking off their study sites and populating the areas with small mammals of four species captured elsewhere and translocated to the study. But they did use low, environmentally realistic application rates of the diazinon. They did not find any difference in mortality between control and treated sites for any of the species after 1 month, but reproduction was negatively affected in several ways. Reproductive condition was reduced from 80% to 20% in exposed males and 100% to 33% in females. Productivity, including percentage of pregnant females and of females giving birth, was significantly reduced in diazinon-exposed animals. The percent of pregnant females ranged from around 14% to 43% in diazinon-exposed animals compared to 40% to 80% in control animals. Similarly, the percent of females giving birth ranged from 0% to 17% in diazinon-exposed animals compared to 22% to 50% in controls. The authors concluded that ecological dynamics in the enclosed prairie grassland ecosystems were disrupted by diazinon through a combination of sublethal reproductive effects, impacting individuals and their populations.

With regard to emigration, there are many studies that have looked at chemical aversion in wildlife. These studies have practical applications in finding repellents that keep wildlife from crops or areas where humans may not want them. Fewer studies have demonstrated area or habitat aversion to contaminants, but it stands to reason that fish and wildlife may be repulsed by sites that stink or have contaminated prey that taste bad. The use of herbicides to remove brush can reduce animal occupancy of an area. The actual reverse, that herbicide use encouraged an influx of animals, was suggested in populations of otters (*Lontra canadensis*) on Vancouver Island, British Columbia. Otters appeared to emigrate from an area of low contamination to one of higher pollution due to better habitat in the polluted site (Guertin et al., 2012), but there were several uncertainties in the study. Also, there are many factors that can cause an animal to emigrate. For example, nest failure in 1 year that was produced by insecticides affecting chicks may cause birds to go to another area the following breeding season. In short, little is known about the direct effects of contaminants on animal emigration or immigration.

9.2.3 Sex Ratios

With the exception of hermaphroditic species that are both male and female, species have only two sexes. At birth or hatching, the male/female ratio is referred to as the **primary sex ratio**. However, there may be differential survival between the sexes—one may be more vulnerable to predation, take more risks, or be more sensitive to toxins than the other, so by the time sexual maturity is reached the numbers of one sex are higher than the other sex. There is also a phenomenon in many species that not all of a given sex breed in a given year. For example, wild stallions monopolize a herd of mares, and larger, more aggressive white-tailed deer bucks mate more often than young bucks, but almost all do mate in a given year. One sex in species of many vertebrates and invertebrates holds territories, and the quality of the territory may attract a greater number of mates so that the male/female breeding success is skewed. Differences between sexes in the number that actually breed are collectively called **operational sex ratios**.

Contaminants may affect one sex more than the other and skew or further skew sex ratios, which may then alter reproduction. In laboratory experiments,

Chen et al. (2015) tested low concentrations of bisphenol A (BPA), used to make plastics and a known endocrine disruptor on zebrafish (*Danio rerio*). Exposure to 0.228 µg/L BPA over two generations resulted in female-biased sex ratio in both F1 and F2 adult populations, decreased sperm density, and decreased sperm quality as measured by motility. Females did not suffer any overt adverse effects, whereas males experienced higher mortality, several types of cellular damage in testes, and impaired spermatogenesis. The authors concluded that continuous exposures to BPA for two generations in zebrafish affects sex ratio and sperm quantity/quality in F1 and F2 adults, impairs reproductive success in offspring from F2 parents, and perturbs various molecular pathways potentially contributing to these BPA-induced male-specific reproductive defects.

Pinzone et al. (2015) examined the persistent organic pollutant (POP) content of long-finned pilot, sperm, and fin whales in the northeastern Mediterranean Sea by sex. They found that all of the POPs were higher in males of the three species than in females. Figure 9.4 only shows total polychlorinated biphenyls (PCBs) and total DDTs, but dieldrin and individual DDTs and PCBs were also higher in males than in females. There were a few statistically significant differences between sexes, but variances within a species and sex were high. In contrast, total hexachlorocyclo-hexane (HCH) was higher in females. Differences among species were attributed to diets. Many of the values and even mean values exceeded the 17 µg/g concentrations considered safe for marine cetaceans (Jepson et al., 1999). Male cetaceans had more individuals with POP concentrations over the values considered to be safe, and their toxic equivalency quotients (TEQs) were higher than that for females, so in these particular concentrations males may be at greater risk than females. The authors attributed the lower concentrations in females to maternal transfer.

In these cases and many others, there were important differences between males and females. In both the zebrafish and whale studies, males were affected more than

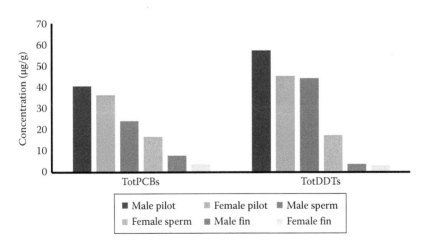

FIGURE 9.4 Concentrations of total PCBs and total DDTs in pilot, sperm, and fin whales by sex. Males are the darker shade, females the lighter. (Adapted from Pinzone, M. et al., *Environ. Res.*, 142, 185, 2015.)

females. While we know that male zebrafish had a higher mortality rate, we are not certain how the elevated POP levels affected male whales. Other studies with other species have demonstrated that female whales are more affected than males, so we cannot draw a broad conclusion here. Nevertheless, numerous studies show a difference between males and females that may be related to skewed sex ratios at some level.

9.2.4 AGE STRUCTURE

Categorizing a population by the ages of its individuals can be very informative about the future status of the population. We can use actual time periods such as years for animals with life spans that encompass several years, or we can use some other way of assessing age such as juveniles (alternatively prereproductives), adults (reproductive), and older (postreproductives) where adults are sexually mature and contribute to the population through breeding. Comparisons among populations within long-lived species can help predict whether a population is likely to increase, decrease, or remain stable through the near future. We can stack the numbers or percent of total population into a pyramid with young prereproductives at the bottom, reproductive next, and postreproductives on top (Figure 9.5). Pyramids with wide bases represent many young relative to other age groups and reflect growing populations. Those with narrow bases and wide tops reflect a population that is becoming senescent and will likely decline with time, and those in between these two extremes reflect stable populations. Care must be taken when comparing age pyramids across species. Some species such as frogs have an intrinsically high mortality rate among young specimens and need a broad base to assure that there will be sufficient young to replace the older generations. Others, such as elephants, have a high survival rate among young and need fewer to replace the adults.

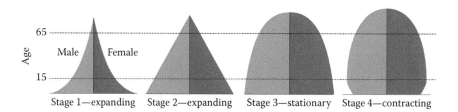

FIGURE 9.5 Age pyramids and what they can reveal. The blue colors are males and the purple females. The two sides do not have to mirror each other as they do in the figures. In stages 1 and 2, the bottom of the pyramid, below age 15, is large relative to the higher portions of the pyramid, and the population is likely to increase in number because there are more young than adults. By stage 3, the base is about even with those under 65 years and we see a situation where the number being born equals the number dying and the population is remaining stable. Stage 4 has a narrow base compared to the other ages and fewer young are coming into the population, which suggests a population that is decreasing. What do you suppose the age pyramid of the human population in the United States more closely resembles? (Courtesy of University of Minnesota Open Libraries, Chapter 1.3, Population and Culture, http://open.lib. umn.edu/worldgeography/chapter/1-3-population-and-culture/.)

Contaminants can alter an age pyramid for a population. There is broad evidence across many taxa and organisms that young individuals are more sensitive than older ones. This may have to do with incompletely formed protective mechanisms such as thinner, more porous dermal layers or less developed cellular mechanisms of defense.

Some scientists have also suggested that being smaller and often less robust than adults, young individuals have more surface-to-volume ratios, which permit greater exposure to chemicals in aquatic species. This is supported by simple mathematics. Let us compare an organism to a cylinder. The equation for the surface area of a cylinder is $2\pi rh + 2\pi r^2$ and that for volume is $\pi r^2 h$, where r is the radius, h is the height, and π is the value of pi. Carry out some calculations for different-sized organisms and you will find that small equates to relatively more surface area. On the other hand, some contaminants are stored in various parts of a body such as lipids and bioconcentrate—these are often higher in older animals and may reenter the circulatory system during periods when fat reserves are used.

As we have mentioned, there are many more types of contaminants in the environment than what we have discussed. In a study using an industrial detergent (Neatex) and a corrosion inhibitor (Norust CR486) found in the delta region of the Niger River in Africa, Ezemonye et al. (2008) found that toxicity differed substantially over a few days' time in young tilapia (*Tilapia guineensis*). The youngest stage tested, 7-day-old fish, was the most sensitive, and by 28 days sensitivity declined by almost 90% for Neatex and 75% for Norust CR486 (Figure 9.6). Both were compounds of a variety of chemicals, so it was not possible to identify a single active agent. The lethal concentrations for the older fish were generally above environmentally realistic values, but the sensitivity of the youngest fish caused concerns.

You might not think that polar bears (*Ursus maritimus*) would be exposed to mercury, but that metal, like many POPs, is ubiquitous throughout the world, even in polar regions. Bechshoft et al. (2016) found that mercury concentrations in polar bear hair in the Hudson Bay of Canada had concentrations that exceeded 7 µg/g, with cubs

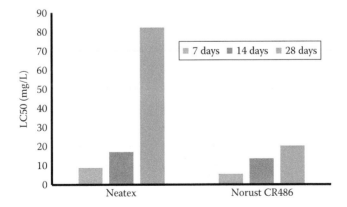

FIGURE 9.6   An example of age sensitivity to two industrial chemicals. In both cases sensitivity (indicated by low LC50 values) is greatest in 7-day-old fish, intermediate in 14-day-old fish, and significantly higher in older fish. (Adapted from Ezemonye, L.I.N. et al., *J. Hazard. Mater.*, 157, 64, 2008.)

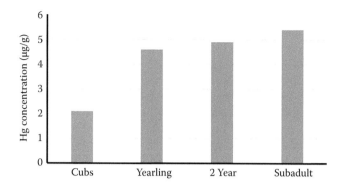

FIGURE 9.7   Whereas young animals often have greater sensitivity to contaminants, they have lower body concentrations of persistent contaminants. In this case young polar bears have less than one half the amount of mercury in their tissues than do the other age classes. (Adapted from Bechshoft, T. et al., *Environ. Sci. Technol.*, 50, 5313, 2016.)

having significantly lower concentrations than other age groups (Figure 9.7). Like many field studies on residues, however, it is not clear if the elevated concentrations of mercury imposed toxicity on the older bears.

We have mentioned maternal transfer where mothers 'download' some of their POPs to their young through lactation or deposition into eggs. In some cases, maternal transfer can lead to higher concentrations of contaminants in the offspring than in the mother. Wang et al. (2012) analyzed the bodies of female harbor seals (*Phoca vitulina*) and their fetuses. The mothers had been harvested for consumption in aboriginal diets. Total HCH was substantially higher in the livers of fetuses than in the mothers while total polychlorinated biphenyl ethers (PCBs) and total poly-brominated diphenyl ethers (PBDEs) were slightly higher in fetuses (Figure 9.8).

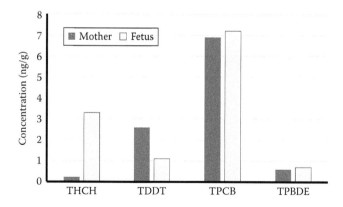

FIGURE 9.8   Due to maternal transfer through the placenta, deposition into eggs, or nursing, young animals can have higher concentrations of some contaminants than their mothers. In this case total HCH, total PCBs, and total PBDEs are higher in harbor seal (*Phoca vitulina*) fetuses than in their mothers. (Adapted from Wang, D.L. et al., *J. Hazard. Mater.*, 223, 72, 2012.)

In contrast, total DDTs were higher in mothers than in fetuses. Other tissues were variable in their relationship between mothers and fetuses. Of 14 possible comparisons between blubber and liver in both age groups, fetuses had higher concentrations than their mothers in 6 instances, but keep in mind any of the POPs found in fetuses had to come from the mother. The authors speculated that these concentrations in fetuses could cause health problems.

9.3   CONTAMINANTS AND LIFE TABLE ANALYSES

Some ecotoxicologists have employed a classical technique of assessing a population's status in evaluating the effects of contaminants. This is a mathematical technique based on the number of organisms living in defined time periods, and through this one constructs a **life table**. Analyses of these life tables help describe (1) population age structure, including peaks and troughs of reproduction; (2) population growth rate; and (3) patterns in survivorship. The presence and absence of contaminants, either through time or between otherwise similar populations, can be superimposed on these life tables to assess if and where effects might appear.

Life tables are built from data gathered during field investigations and focus on the age distribution of populations. They count or estimate the number of individuals at different age classes or stage of development (n) at various time periods (t) based on the longevity of the organisms in the population. For instance, if a scientist was interested in developing a life table for a species of insect, she might look at the egg, larva, pupa, and adult and perhaps sample weekly or more often. If she was studying hippopotami, she might mark the animals to determine their age in years and sample annually. Sometimes scientists also have access to reproductive data such as how many young each age class produces. These values can provide supplemental information to further describe a population. Ecologists use these tables to help elucidate the factors and processes that govern population growth. Ecotoxicologists can extend the application of life table analysis to characterize chemical effects on populations. As we have stated previously, but bears repeating, contaminants in this context can be viewed as one of the many factors of perturbations that influence a population's dynamics.

There are two types of life tables. In one, a segment of a population called a cohort is counted and followed from a given start time to death. This leads to *cohort life tables* because one is following a cohort or group of organisms; an alternative name for these is *dynamic life tables*. Such dynamic life tables are most often applied to short-lived organisms or long-lived biota amenable to longitudinal studies with surveys conducted every few years. As an alternative to dynamic life table analysis, *static life table* or time-specific table is made from data collected during one time period and partitioned into the various age categories. Static life table analysis is less frequently followed in field surveys than cohort tables.

As an illustration of cohort life table analysis, Table 9.1 summarizes the output of a typical life table. The math behind this table is as follows:

x = age in years of stage (egg, larvae, nymph, adult or various instars, pupa adult) or year of study

n_x = number of individuals in each age or stage x

TABLE 9.1
Hypothetical Example of a Life Table and Its Calculations

Age	n_x	l_x	λ	d_x	q_x	b_x	$l_x b_x$	$\sum x l_x b_x$
0	500	1.00	0.65	0.35	0.35	0	0	0
1	325	0.65	0.89	0.07	0.11	0.10	0.06	0.065
2	290	0.58	0.96	−0.02	−0.03	0.50	0.29	0.64
3	300	0.60	1.03	−0.01	−0.02	4.0	2.4	7.8
4	295	0.59	0.98	−0.03	−0.05	5.0	2.9	19.7
5	310	0.62	1.05	0.12	0.19	1.0	0.62	22.2
6	200	0.40	0.64	0.20	0.50	0.5	0.20	24.0
7	100	0.20	0.50	0	0	0	0	

Note: $R_0 = 6.5$ offspring; $T = 3.6$ years.

l_x = percent of the original cohort that survives to age or stage x (= n_x/n_0 where n_0 is the number living in the first age group)

d_x = probability of dying during age or stage x (= $l_x - l_{x+1}$). If dx is negative, then either immigration or birth exceeded deaths in stage x compared to stage x + 1

q_x = percent of dying between age or stage x and age or stage x + 1 (= d_x/l_x)

b_x = number of offspring produced per individual in age/stage x. bx is not part of the traditional life table but is added here to provide further detail.

These data can be used to calculate additional variables including:

λ or the finite rate of increase at any increment. This is measured as the ratio of population size at the end of one interval to population size at the end of the previous interval.

$$\lambda = \frac{n_{x+1}}{n_x}.$$

When $0 < \lambda < 1$, population decreases during that increment.

$\lambda = 1$, population is stable.

$\lambda > 1$, population increases. λ is useful for an open life table that accounts for immigration and birth into a population along with the factors of emigration and death.

R_0, or net reproductive rate, where $R_0 = \Sigma(l_x b_x)$.

T, the generation time (the time between the birth of one cohort and the birth of their offspring), where $T = (\Sigma x l_x b_x)/R_0$.

r, the per capita rate of increase (as is R_0, r is a measure of the change in population size) and is calculated as $r = \ln R_0/T$.

Note that R_0 and r do not estimate the same population parameter; r tells us whether a population is experiencing more births than deaths, but R_0 tells us if the population is increasing or decreasing relative to some fixed point defined as year 1. In other words

If $R_0 > 1$, the population is increasing.
If $R_0 < 1$, the population is decreasing.
If $R_0 = 1$, the population is constant.
If $r > 0$, then there are more births than deaths.
If $r < 0$, there are more deaths.
If $r = 0$, births and deaths are equal.

In one particular example $R_0 = 6.52$, $r = 0.57$, T or generation time is 3.7 years.

In this case, R_0 and r tell us that the population is increasing. Suppose that these data came from a reference population, one could compare the R_0 and r values to another population in an area that is contaminated. If the R_0 or r values indicate a declining population or a growth rate that is substantially lower than that of the reference population, one might conjecture that the contaminants are having a population effect. Keep in mind, however, that other stress factors might be involved, so the selection of a reference area that is as close to the contaminated site as possible is critical.

Sha et al. (2015) used life tables to study the effects of PBDEs on the marine rotifer *Brachionus plicatilis*. They used laboratory populations and had data on age-specific reproduction. The authors determined rotifer population demographic parameters including age-specific survivorship (l_x), age-specific fecundity (m_x), net reproductive rate (R_0), intrinsic rate of increase (r), finite rate of increase (λ), life expectancy (E_0), and generation time (T) and used them as measures of treatment effects. This study showed increasingly intense negative effects on many of the rotifer demographic parameters with elevated PBDE concentrations. The population growth curves of control animals showed almost instantaneous growth and reached peak abundances within 11 days, but those exposed to PBDE-209 did not show any increase in numbers until after 5 days. Also, the time for population growth to level off was reduced by PBDEs compared with controls. PBDE exposure also reduced peak population densities. The authors concluded that life table demography and the population growth curve can be used to evaluate PBDE effects.

Martinez-Jeronimo et al. (2013) exposed cohorts of recently hatched copepods, *Daphnia schoedleri*, to three sublethal concentrations (0.54, 5.4, and 54 ng/L) of the pyrethroid α-cypermethrin plus a control for 21 days. *Daphnia schoedleri* is another species that can be easily maintained in the laboratory for many generations. Effects were measured through a life table analysis for fecundity (m_x) and survivorship (l_x). In addition, the intrinsic rate of population growth (r), net reproductive rate (R_0), life expectancy at birth (E), generation time (T), and the average life span were calculated in each cohort. Survivorship curves showed substantial loss to the populations even at 0.54 ng/L α-cypermethrin. Moreover, all life table parameters were better in the control group than any of the treatments. The α-cypermethrin negatively affected the average life span, life expectancy at hatching, net reproductive rate, intrinsic rate of population growth, and generation time at the highest concentration.

One final example should suffice to convey the utility of this tool. Levin et al. (1996) exposed polychaete worms *Capitella* sp. and *Streblospio benedicti* to contaminants often found in estuaries including sewage, blue-green algae, fuel oil, and a control. Blue-green algae are not really contaminants in the strict sense, but they form blooms in response to nutrient enrichment and often produce toxins. The authors then collected and analyzed data in a life table context to compare the demographic responses of both species to these forms of pollution. Separate cohorts consisted of each species in each treatment. For both species, survival was high in all treatments, except the blue-green algae treatment where oxygen depletion occurred. However, treatments obviously affected age of maturity, fertility, and generation time between species and among contaminants. Population growth rates (r) were higher in *Capitella* sp. than in *S. benedicti* for all treatments, primarily due to earlier maturation and greater fertility in *Capitella* early in the study. For *Capitella* sp., sharp increases in r occurred in the sewage and blue-green algae relative to controls and fuel oil treatments. This marked increase was primarily due to reduced maturation time and increased age-specific fertility. Fuel oil reduced r through delaying maturation and reducing age-specific fertility. Population growth rates of *S. benedicti* in the fuel oil and algae treatments were lower than that in control and sewage treatments. Fuel oil delayed maturity and reduced fertility in this species while the blue-green algae reduced survival of juveniles and fertility of adults. *Capitella* was less sensitive to these forms of estuarine pollution than *S. benedicti*. For both species, all demographically important effects of contaminants occurred early in life, suggesting a need to focus on juveniles and young adults in field and laboratory testing.

In summary, contaminants can have significant effects on population status. These effects can be observed by using the same parameters that ecologists use to characterize populations in field studies.

9.4   CHAPTER SUMMARY

1. It can be very difficult to demonstrate conclusively that a contaminant or set of contaminants actually depress populations in the field. There are many other sources of stress that have to be controlled or accounted for before single factors can be proven to be responsible for substantial mortality. Even with demonstrated mortality, it must be shown that the contaminants are acting in an additive rather than a compensatory manner. Making proof even more difficult, contaminants at environmentally realistic concentrations are more likely to induce subtle sublethal effects such as lethargy, reduced awareness, or reduced reproductive output than overt acute mortality.

2. Like other stress factors operating on a population, contaminants can be either density dependent or independent, at least in theory, depending on conditions. In one way, contaminants might be considered density dependent if different concentrations of the chemical are considered. However, it is more likely under typical field conditions for contaminants operating in a density independent fashion.

3. Abundance is a key characteristic of a population and most often expressed as density, the number of organisms per unit area. Population growth is simply the change in abundance over time. Growth in particular can be expressed as changes in the general equation:

Growth = Natality rate − Mortality rate + Immigration rate − Emigration rate

Substantial data exists on the effects of contaminants on birth and death rates but not on immigration or emigration, although arguments can be raised on how they might influence these last two parameters.

4. Sex ratios can be influenced by contaminants, as several studies have shown, on a diverse assemblage of species. It is possible that contaminants may have a differential toxicity to males and females. Endocrine disruptors may alter the operational sex ratios of populations. Differential uptake and accumulation of contaminants between males and females are widely known, although the effects of this accumulation are not always clearly understood.

5. Contaminants can alter the age structure of populations. Often smaller, younger individuals are more sensitive to chemical exposures than older members of the population. Older individuals, however, may have had more time to accumulate contaminants, especially those that are lipid soluble and bioconcentrate. Maternal transfer may dump contaminant loads into developing young.

6. Life table analysis is one way of assessing if contaminants may be affecting a population. By collecting data on abundance and reproductive information either through time or at a single time, ecotoxicologists can compare populations in reference areas (or controlled laboratory situations) with those in contaminated sites and assess if contaminant exposure is producing harm to populations.

9.5 SELF-TEST

1. Short answer. Why is it difficult to prove that a contaminant or a group of contaminants is having an effect on a population density?

2. Which of the following is true about compensatory mortality?
   a. It is always density dependent.
   b. Mortality caused by one factor would be caused by another in the absence of the first factor.
   c. It compensates for populations that have exceeded their carrying capacity.
   d. More than one above.

3. At a population level of 300 animals, contaminants kill 50 individuals; when the population reaches 450, 75 animals die; and when the population is 600 animals, contaminants kill 100 individuals. This would be an example of
   a. Density dependence
   b. Density independence
   c. Additive mortality
   d. Cannot tell by the information provided

4. On what element of population growth do we have the most data for contaminant-related effects?
   a. Natality
   b. Mortality
   c. Emigration
   d. Immigration
5. True or False. Young animals are often more sensitive to environmental contaminants that older, sexually mature members of the same species.
6. True or False. Metam sodium proved to be more toxic to snapping turtle embryos than several pesticides that were used at the same time.
7. In what ways can differences occur between males and females of the same species with regard to the effects of contaminants?
   a. Females can discharge some of their contaminant load through maternal transfer.
   b. One sex may have more fat deposits than the other and have higher concentrations of lipid-soluble contaminants.
   c. Endocrine disruptors may affect males differently than females.
   d. All of the above.
8. Short answer. What is an age pyramid and how can it help predict whether a population is increasing, decreasing, or remaining stable?
9. True or False. No studies have shown that contaminants can decrease the intrinsic rate of increase (r) when compared to reference populations.
10. What life table parameter tells us if a population is increasing, decreasing, or remaining the same?
    a. b_x
    b. l_x
    c. q_x
    d. R_0

REFERENCES

Bechshoft, T., Derocher, A.E., Richardson, E., Lunn, N.J., St. Louis, V.L. 2016. Hair mercury concentrations in Western Hudson Bay polar bear family groups. *Environ Sci Technol* 50: 5313–5319.

Burnham, K.P., Anderson, D.R. 1984. Tests of compensatory vs additive hypotheses of mortality in mallards. *Ecology* 65: 105–112.

Chen, J.F., Xiao, Y.Y., Gai, Z.X., Li, R., Zhu, Z.X. et al. 2015. Reproductive toxicity of low level bisphenol A exposures in a two-generation zebrafish assay: Evidence of male-specific effects. *Aquat Toxicol* 169: 204–214.

de Solla, S.R., Palonen, K.E., Martin, P.A. 2014. Toxicity of pesticides associated with potato production, including soil fumigants, to snapping turtle eggs (*Chelydra serpentina*). *Environ Toxicol Chem* 33: 102–106.

Ezemonye, L.I.N., Ogeleka, D.F., Okieimen, F.E. 2008. Lethal toxicity of industrial chemicals to early life stages of *Tilapia guineensis*. *J Hazard Mater* 157: 64–68.

Guertin, D.A., Ben-David, M., Harestad, A.S., Elliott, J.E. 2012. Fecal genotyping reveals demographic variation in river otters inhabiting a contaminated environment. *J Wildl Manage* 76: 1540–1550.

Jepson, P.D., Bennett, P.M., Allchin, C.R., Law, R.J., Kuiken, T. et al. 1999. Investigating potential associations between chronic exposure to polychlorinated biphenyls and infectious disease mortality in harbor porpoises from England and Wales. *Sci Total Environ* 243(244): 339–348.

Levin, L., Caswell, H., Bridges, T., DiBacco, C., Cabrera, D., Plaia, G. 1996. Demographic responses of estuarine polychaetes to pollutants: Life table response experiments. *Ecol Appl* 6: 1295–1313.

Martinez-Jeronimo, F., Arzate-Cardena, M., Ortiz-Butron, R. 2013. Linking sub-individual and population level toxicity effects in *Daphnia schoedleri* (Cladocera: Anomopoda) exposed to sublethal concentrations of the pesticide alpha-cypermethrin. *Ecotoxicology* 22: 985–995.

Pinzone, M., Budzinski, H., Tasciotti, A., Ody, D., Lepoint, G. et al. 2015. POPs in free-ranging pilot whales, sperm whales and fin whales from the Mediterranean Sea: Influence of biological and ecological factors. *Environ Res* 142: 185–196.

Sha, J., Wang, Y., Lu, J., Wang, H., Chen, H. et al. 2015. Effects of two polybrominated diphenyl ethers (BDE-47, BDE-209) on the swimming behavior, population growth and reproduction of the rotifer *Brachionus plicatilis*. *J Environ Sci* 28: 54–63.

Sheffield, S.R., Lochmiller, R.L. 2001. Effects of field exposure to diazinon on small mammals inhabiting a semi-enclosed prairie grassland ecosystem: 1. Ecological and reproductive effects. *Environ Toxicol Chem* 20: 284–296.

Sparling, D.W., Bickham, J., Cowman, D., Fellers, G.M., Lacher, T. et al. 2015. In situ effects of pesticides on amphibians in the Sierra Nevada. *Ecotoxicology* 24: 262–278.

Sparling, D.W., Fellers, G.M., McConnell, L.L. 2001. Pesticides and amphibian population declines in California, USA. *Environ Toxicol Chem* 20: 1591–1595.

Stansley, W., Roscoe, D.E. 1999. Chlordane poisoning of birds in New Jersey, USA. *Environ Toxicol Chem* 18: 2095–2099.

Wang, D.L., Atkinson, S., Hoover-Miller, A., Sheiver, W.L., Li, Q.X. 2012. Organic halogenated contaminants in mother-fetus pairs of harbor seals (*Phoca vitulina richardii*) from Alaska, 2000–2002. *J Hazard Mater* 223: 72–78.

10 How Contaminants Can Affect Community and Ecosystem Dynamics

10.1 INTRODUCTION

When we start understanding how contaminants can affect higher levels of ecological organization such as communities and ecosystems, we are looking at a matter of scale and realizing that what happens at much lower levels can influence higher levels. In a previous publication (Sparling, 2016), Dr. Greg Linder clearly illustrated this with a diagram, which we include here as Figure 10.1. At the lowest level of organization, we have the cell and its components, which can be physiologically poisoned by exposure to all sorts of contaminants. At this level we may be interested in cellular metabolism or cell death called **apoptosis**. The changes that occur at this level may be sufficiently strong to affect organs and organ systems through cancer or other malfunctions of these systems. Given sufficient toxicity, the organism may suffer reduced reproduction, premature mortality, or other toxicity. If enough individuals of a species are affected, there may be population effects through altered sex ratios, reduced fertility rates, altered overall numbers, or increased mortality rates that reduce population size or even lead to local extinction, otherwise called **extirpation**. Because virtually all of the species in an area interact with each other through food chains, decomposition, and various forms of symbioses, what affects one species may ultimately have ramifications on other species in the community. These perturbations can then lead to altered energy relationships or imbalances in nutrient and elemental cycles that are part of ecosystems.

Of these various levels, contaminant effects at the suborganism and organismal strata are most easily identified because, in many cases, controlled experiments can be conducted under laboratory conditions. At higher levels, effects may be more difficult to conclusively identify, as we have seen with populations in the previous chapter. Clear, unmitigated proof of contaminant effects at community and ecosystem levels is often more difficult to establish than for populations, but there are several studies that have tried to do exactly that and their findings are the basis for this chapter.

10.2 ASPECTS OF COMMUNITY ECOLOGY

10.2.1 Species Richness, Diversity, and Abundance

Before we discuss how contaminants affect communities, it might be a good idea to present some ideas of what we mean by communities. A **community** is a group of

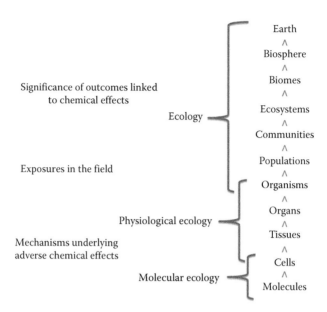

FIGURE 10.1   An explanatory model depicting the scale or ecotoxicological organization from intracellular events to the biosphere of the earth. (From Sparling, D.W., *Ecotoxicology Essentials*, Academic Press, Boston, MA, 2016; *Credit:* Dr. Greg Linder. With permission.)

species living and interacting or potentially interacting within an area. These species include all the microbes, fungi, plants, and animals in an area, which might consist of a wetland, a terrestrial landscape, or even an ocean. A fundamental characteristic of a community is **species richness** or the number of different species occupying that habitat. **Species diversity**, a related topic, takes into account the species richness of an area and how equal those species are in numbers. For example, a simple community with a species richness of five would have greater diversity if there was a high level of **evenness** in the numbers of each species, for example, they each may have 100 members. If there was wide variation in numbers, such as when one species has 500 members and the others 100 or less, diversity declines and one species may become dominant. High species diversity and richness generally reflect healthy, stable, energetically efficient communities, whereas communities that are heavily dominated by one species are less likely to be stable because if something should harm the dominant species there may be a gap in energy or nutrient exchange. Contaminants can enter into the equation when they start decimating sensitive species, leaving only the most tolerant species and reducing diversity.

There are many studies supporting the notion that contaminants can reduce species diversity and richness. Often this is due to variability in sensitivity among species to a given chemical or set of chemicals. Note also that studies that do not show a reduction in species diversity are also common. No doubt the differences lie in a complex relationship among species, contaminants, and concentrations of those contaminants. In one study that did show a decrease in diversity, Baguley et al. (2015) examined the marine meiofauna communities in 66 marine benthic stations in the

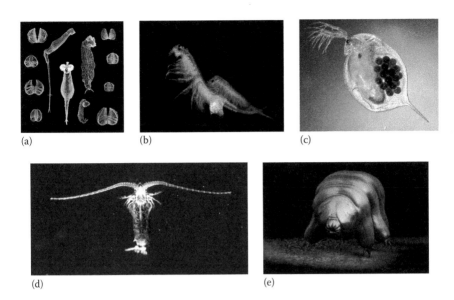

FIGURE 10.2 Examples of meiofauna, rotifers: (a) rotifers, (b) crustacea, (c) cladoceran, (d) copepod, and (e) Tardigrada. (a: Courtesy of Gross, L., Diego Fontaneto—Who needs sex (or males) anyway?, *PLoS Biol.,* 5(4), e99, 2007, doi: 10.1371/journal.pbio.0050099; b: Courtesy of Hans Hillawaert, https://en.wikipedia.org/wiki/Brine_shrimp#/media/File:Artemia_salina_4. jpg; c: Ebert, D., Own work, CC BY-SA 4.0, Basel, Switzerland, https://commons.wikimedia. org/w/index.php?curid=47132022; d: Courtesy of http://www.montereybayaquarium.org/ storage/animals/520x260/Copepod.jpg.)

Gulf of Mexico that potentially had been exposed to oil from the Deepwater Horizon blowout and oil spill of 2010. Meiofauna are relatively sedentary, small multicellular animals that are larger than bacteria and include nematodes, copepods, and the like (Figure 10.2). As a group they are very important ecologically because they form the basis for food chains and energy pyramids.

Baguley et al. (2015) found that their abundance, diversity, and the nematode-to-copepod ratio (N:C) varied significantly across impact zones. Nematodes became more dominant with increasing oil impacts and spiked near the actual wellhead. With nematode dominance, major taxonomic diversity and evenness decreased in more polluted zones close to the wellhead. The abundance of copepods and other minor meiofauna decreased in the most severely impacted zones, and some taxa were virtually wiped out.

10.2.2 Food Chains and Webs

Another aspect of community ecology that can be strongly influenced by contaminants is food webs or chains. The difference between a food web (Figure 10.3) and a chain is that a web is more complex and usually more reflective of actual conditions, except in very simple environments such as the Arctic. Food webs/chains have been extensively documented in transporting lipophilic contaminants that bioconcentrate

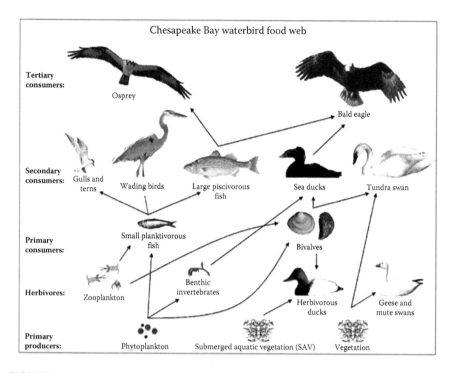

FIGURE 10.3   Illustration of a food web in the Chesapeake Bay. (Courtesy of Perry, M.C., Changes in food and habitats of waterbirds, Chapter 14, pp. 60–62, in: Phillips, S.W. (ed.), *Synthesis of U.S. Geological Survey Science for the Chesapeake Bay Ecosystem and Implications for Environmental Management*, U.S. Geological Survey Circular 1316, 63pp.)

from lower levels to higher ones, even top levels such as from zooplankton to fish-eating birds described in Chapter 7 for DDT. As we mentioned there, biomagnification can increase tissue concentrations by a million-fold over water concentrations. Biomagnification up a food web is not limited to organochlorine pesticides; however, many organic pollutants have that ability.

In a bay of the South China Sea, Zheng et al. (2016) collected various samples ranging from water to sediment to organisms and found that total polybrominated diphenyl ethers (PBDEs) tended to increase as one progressed up the food web (Figure 10.4). Individual ethers were not as predictable as total concentration, however. The relatively low total PBDEs in wolffish reflect that the species preys extensively on crabs and similar species rather than on cuttlefish, which is also a predator. Food webs are also a way for contaminants to move from one ecological compartment to another as when fish-eating birds or mink consume contaminated aquatic prey. Removal of sensitive keystone species from these webs through contaminant toxicity may disrupt entire food webs. **Keystone species** are those that contribute more to the ecology of a community than would be predicted by biomass alone and often include selected species at or near the bottom of a food web that convert water- or sediment-borne contaminants into biological tissues.

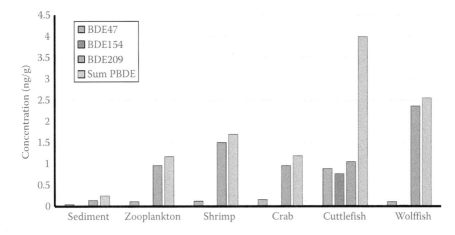

FIGURE 10.4 Bioaccumulation characteristics of polybrominated diphenyl ethers in the marine food web of Bohai Bay. Note that most of the contaminants are higher in organisms than in the sediment. Total PCBs appear to have the strongest ability to biomagnify. (Adapted from Zheng, B.H. et al., *Chemosphere*, 150, 424, 2016.)

10.2.3 SYMBIOTIC RELATIONSHIPS

Various forms of symbiotic (literally 'living together') relationships can occur between species in a community. These can include predator–prey, parasite/disease–host, competition, mutual relationships in which both species benefit or there are one-sided beneficial, or detrimental partnerships. Contaminant dynamics may be altered by such interactions. For example, Wang et al. (2016) identified complex relationships between corn (*Zea mays*), the tiny root-like arbuscular mycorrhizae (AM) of fungi, and zinc oxide (ZnO) nanoparticles. Arbuscular mycorrhizae are fungal filaments that penetrate the roots of many different kinds of vascular plants and derive nutrition from the plants and, in turn, help capture nutrients such as phosphorus, sulfur, nitrogen, and micronutrients from the soil that benefit the plants. The nanoparticles had no significant adverse effects on corn or hyphae at a concentration of 400 mg/kg but inhibited corn growth at concentrations at and above 800 mg/kg. At these higher levels, ZnO diminished the ability of plants to acquire nutrients and decreased photosynthetic pigment concentration and root activity of the plants. The nanoparticles also increased zinc concentrations in the plants. However, AM inoculation significantly reduced the negative effects of the nanoparticles, and inoculated plants showed increased growth, nutrient uptake, and photosynthetic pigment content and decreased zinc concentrations compared to plants without inoculation. The authors concluded that at high contaminations ZnO nanoparticles caused toxicity to both plants and fungi but that AM inoculations helped mitigate zinc toxicity by decreasing the availability of the metal to the plants. Other organisms such as some strains of bacteria, microalgae, and several endophytic fungi convey similar protection to vascular plants. Interestingly, some contaminants are more toxic to the mycorrhizae than to the plants they protect and function to reduce the vigor of these defense systems.

In a more complex example, Gustafson et al. (2016) examined the relationship between the herbicide atrazine with two types of parasites and their hosts. Trematode parasites are common in aquatic environments and can cause diseases in humans, livestock, and wild animals. The researchers used a trematode that uses a snail as a first intermediate host. When it emerges from the snail, it continues its life cycle in a secondary intermediate host of an ostracod, a representative of meiofauna, and it becomes an egg-laying adult in amphibians. Their results demonstrated that atrazine impairs trematode transmission by altering snail and ostracod host–parasite interactions. Although atrazine did not affect the survival of uninfected snails, it reduced the life span of infected snails. As a result, the number of trematode cercaria (a larval stage) produced by snails was half of that of controls at 3 µg/L of atrazine and 15% at 30 µg/L. Atrazine directly killed uninfected ostracods at 30 µg/L, but the presence of trematodes decreased the vulnerability of ostracods to the herbicide. At the higher dose level, the number of trematode metacercariae (a more advanced developmental stage than cercaria) infections declined to 32% of controls at the low dose and 29% at the higher dose. The combination of reduced cercaria production and reduced metacercarial infection in the 3 and 30 µg/L atrazine treatment groups reduced the net number of infective worms produced to 16.4% and 4.3%, respectively, relative to the control. These results demonstrate a complex nature of pesticide effects on parasitism and indicate that atrazine had different effects on the two life stages of trematodes. The authors did not examine parasitism of amphibians to complete this scenario, but it would not be surprising if their rate of parasitism also declined.

10.2.4 ECOLOGICAL SUCCESSION

Another aspect of community ecology is **ecological succession**. This term is most often applied to plant communities, but it can also be meaningful for other organisms such as microbes. It is the gradual establishment or replacement of a community after a significant destructive event. Primary succession (Figure 10.5) occurs when there is no or very little of a community prior to the event and can occur on barren ground as plants or microbes colonize the space. Secondary succession takes place after a disturbance when a base of plant life still exists. Secondary succession occurs more rapidly than primary succession because soil formation is one of the most time-consuming processes in succession, and secondary succession takes place on soils that still exist with a seed or spore bank left. Secondary succession often results in a community very similar to what had been there prior to the disturbance.

Oil spills can have major effects on plant communities along shorelines, estuaries, and mangrove stands and are one of the many ecological perturbations that can occur in these habitats. Anthropogenic pressures alter natural ecosystems and the ecosystems are not considered to have recovered unless secondary succession has returned the community to something resembling the preexisting state. However, depending upon time, space, and intensity of disturbance, succession may (1) follow natural restoration through secondary succession, (2) be redirected through ecological restoration, or (3) be unattainable (Borja et al., 2010).

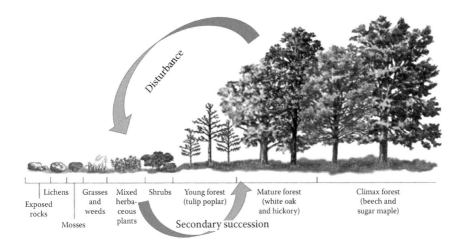

Lichens | Grasses and weeds | Mixed herbaceous plants | Shrubs | Young forest (tulip poplar) | Mature forest (white oak and hickory) | Climax forest (beech and sugar maple)

Exposed rocks

Mosses

Secondary succession

FIGURE 10.5 Illustration of primary and secondary succession. Primary succession starts with exposed soil and rocks and eventually becomes a forest; secondary succession occurs with primary succession is set back and subsequently becomes a forest again, as indicated by the arrow. (Courtesy of Pinterest and Chen, P., *Botany 320: Prairie Ecology*, College of Du Page, 2012, http://bot1320.nicerweb.com/Locked/media/ch10/succession.html.)

The relatively recent Deepwater Horizon oil spill of 2010 can serve as an example of the effects on secondary succession. In moderately oiled marshes, total petroleum hydrocarbon concentrations in surface soils were in the 70 mg/g range (Lin et al., 2016). There were signs of ecological damage but the biomass of the common marsh plants *Spartina alterniflora* and *Juncus roemerianus* were similar to that in reference areas within 24–30 months post spill. In contrast, in heavily contaminated sites, hydrocarbon concentrations in surface soils exceeded 500 mg/g 9 months after the spill and marsh plants did not fully recover. Heavy contamination resulted in a nearly complete die-off of plants, and even after 42 months, the biomass of the two plant species was only 50% of the reference values. Heavy oiling also changed the vegetation structure of shoreline marshes from a mixed *Spartina–Juncus* community to predominantly *Spartina*. Although the living biomass of *Spartina* aboveground recovered within 2–3 years, *Juncus* showed no recovery.

For evidence that a variety of contaminants can influence ecological succession, we can turn to the work by Olson and Fletcher (2000) who evaluated a former industrial sludge basin containing organic pollutants. The basin had undergone ecological recovery over a 13-year period, following the removal of surface water. The investigators found that typical phases of ecological recovery including plant invasions and succession had occurred but that the structure and diversity of the plant community was different from that at a nonpolluted but disturbed site. Three plant species believed to be the early invaders of the basin still persisted in large numbers, indicating that these species could cope well with environmental stresses at this area and with organic pollutants. Clearly, succession had been slowed due to the presence of polycyclic aromatic hydrocarbons still remaining in the soil.

Often land users such as agricultural producers want to set succession back or maintain a community at a specific successional stage such as a grassland. Herbicides are often the mechanism of choice to do this. In contrast, wildlife managers who also want to maintain a stand at a particular successional stage because of the wildlife it attracts often use mechanical means such as burning or cutting and rely on herbicides to a lesser extent.

10.2.5 COMMUNITY SENSITIVITY

An interesting study was conducted by Bendis and Relyea (2016) that showed the sensitivity of communities to contaminants and the importance of keystone species. As we mentioned above, small aquatic invertebrate species such as copepods and amphipods can often function as keystone species because they form the basis for food chains and start the process of incorporating water-borne contaminants into the biological component. Agricultural pesticides such as glyphosate, atrazine, and a host of other chemicals can run off into wetlands and affect their ecology. With prolonged exposure to these pesticides, organisms can develop a tolerance and be able to survive in polluted environments. Bendis and Relyea (2016) identified two populations of the copepod *Daphnia pulex*—one that had developed a resistance to the insecticide chlorpyrifos and one that had not because it came from a source distant from agricultural fields. They conducted an extensive outdoor mesocosm experiment with the individual communities furnished with phytoplankton, periphyton, and leopard frog (*Rana* [*Lithobates*] *pipiens*) tadpoles. They then added one or the other *Daphnia pulex* populations to each mesocosm. The communities were subsequently exposed to (1) no insecticide; (2) one of three different concentrations of the AChE-inhibiting insecticides chlorpyrifos, malathion, or carbaryl; or (3) pyrethroid insecticides that had a different mechanism of effect than the AChE inhibitors. Communities containing sensitive *D. pulex* and moderate to high concentrations of insecticides (and hence had affected populations of *D. pulex*) experienced phytoplankton blooms and negative effects through all trophic groups including amphibians. However, communities containing resistant *D. pulex* were buffered from these effects at low to moderate concentrations of all AChE-inhibiting insecticides. This showed that *D. pulex* had developed a cross-resistance to all three of the AChE-inhibiting insecticides. They were not buffered against pyrethroid insecticides, however, and these communities showed similar responses to those with sensitive *D. pulex* populations and AChE inhibitors. The authors concluded that a simple change in the population-level resistance of zooplankton to a single insecticide can have widespread consequences for community stability and that the effects can be extrapolated to other pesticides that share the same mode of action.

10.3 CONTAMINANTS AND ECOSYSTEMS

When ecologists study ecosystems, they focus on the interactions between the biotic portion of an area (individuals, population, and communities) and the abiotic elements (energy; chemicals; nonliving components of soil, water, and air; weather; climate, etc.). Common energy measurements of ecosystem dynamics include **respiration (R)**, often measured as the amount of carbon dioxide produced or the amount of oxygen used during the carbon dioxide cycle and is related to energy consumption by the ecosystem; **photosynthesis (P)** (P may also stand for productivity when animals are being investigated)—often measured as the amount of carbon dioxide taken in or by the amount of carbon dioxide assimilated; and the **P/R ratio**. The P/R ratio is important for if its value is <1, the system is an energy drain; when P/R = 1, the system is in energy homeostasis; and when P/R > 1, the system is generating more energy than it uses. The amount of sunlight penetrating the system is another factor for sunlight serves as the principal source of external energy. **Productivity**, the rate of biomass an ecosystem generates, is another factor that belongs to ecosystem ecology. While productivity measures rate, **production** quantifies amount, as in kilograms or bushels. **Primary productivity** is the rate of biomass generated by green plants, and **secondary productivity** is the rate of biomass generated by animals.

Ecosystem ecology may also focus on **nutrient and mineral balances** between soil and organisms. Soils may be naturally low in required nutrients or they may have toxic levels of certain minerals such as salt, both of which can alter productivity. In this regard, ecologists quantify elements of nutrient cycles such as the carbon cycle or nitrogen cycle (Figure 10.6) and water balance. Questions asked during such investigations may include the following: (1) How does the natural chemistry of the environment affect the organisms that live there? (2) How is energy channeled through the various compartments including organisms, air, water, and soil/sediment? (3) How are nutrients cycled through these compartments? Because contaminants are chemicals,

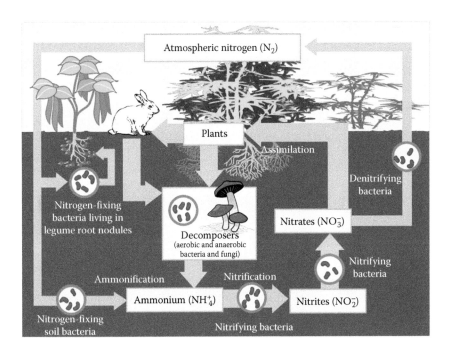

FIGURE 10.6 The nitrogen cycle—one of several mineral and gas cycles. The atmosphere serves as the primary reservoir for nitrogen. It can be assimilated through leaves and stems of plants, or be converted into other nitrogen forms by bacteria. Eventually, soil nitrogen can be recycled into the atmosphere by denitrifying bacteria. (Courtesy of Wikipedia: U.S. EPA, Nitrogen cycle, 2015, https://www3.epa.gov/caddis/ssr_amm_nitrogen_cycle_popup.html.)

they justifiably belong as elements of ecosystem science. A difference between ecologists and ecotoxicologists in this context is that ecologists focus more on natural chemicals whereas ecotoxicologists are mostly interested in anthropogenic chemicals.

One quick example of how contaminants interact with ecosystem ecology is seen in the Bendis and Relyea (2016) study cited above. In addition to changes in the communities, the authors also observed significant energy and chemical effects related to treatments. Temperature in the mesocosms was mildly affected by the treatments. Of course, it could be argued that variability due to light–dark cycles and prevailing temperatures in outdoor mesocosms may have obscured any significant differences. Both insecticide treatment and *D. pulex* populations significantly affected water pH, with the mix of AChE-inhibiting pesticides and sensitive *D. pulex* populations leading to increased pH whereas AChE-inhibiting pesticides and resistant *D. pulex* having lower pH values. Dissolved oxygen followed similar trends, and light attenuation generally increased with higher concentrations of all insecticides.

In an extensive review of the effects of contaminants on ecosystem functioning in marine and estuarine habitats, Johnston et al. (2015) found that toxic contaminants generally reduced productivity and increased respiration. Effects varied according to the type of contaminant and the component(s) of the system that were studied (e.g., particular trophic levels, functional groups, or taxonomic groups). Most of the studies reviewed by Johnston et al. (2015) dealt with planktonic communities instead of total biodiversity. Unfortunately, the studies rarely interpreted their findings in terms of ecosystem ecology. Most studies identified negative impacts of contaminants on primary production or productivity. The author stated that an understanding of chemical contaminant effects will remain patchy until further studies focus on factors associated with ecosystem ecology. They also suggested that productivity and respiration may serve as key endpoints in such studies.

One of the major factors potentially affecting ecosystem functioning across the world is global climate change. A plethora of studies and reports have warned that climate change may alter many food webs, energy transfer ratios, nutrient cycling, and just about everything one can think of ecologically. Conceptually, there are two elements to climate change. One is the natural change in climate that occurs all of the time but may trend in one direction or another for a period. The other element is that caused by anthropogenic pollution, particularly the so-called greenhouse gases of carbon dioxide, methane, nitrous oxide, and water vapor that are produced through the combustion of fossil fuels and other sources. A vast majority of climate scientists strongly support that anthropogenic pollution is a contributing factor, perhaps even the main factor, to climate change. In its fifth assessment of global climate change, the Intergovernmental Panel on Climate Change, an international body of climate experts, stated (IPCC, 2014) the following:

> Anthropogenic greenhouse gas emissions have increased since the pre-industrial era, driven largely by economic and population growth, and are now higher than ever. This has led to atmospheric concentrations of carbon dioxide, methane and nitrous oxide that are unprecedented in at least the last 800,000 years. Their effects, together with those of other anthropogenic drivers, have been detected throughout the climate system and are extremely likely to have been the dominant cause of the observed warming since the mid-20th century.

Polar bears (*Ursus maritimus*) have been a 'poster child' for the threats of global climate warming for their very habitat, polar ice, is diminishing at an alarming rate. In one study on polar bears, McKinney et al. (2009) showed that climate change is affecting polar bear diets and that these changes are increasing the bears' uptake of persistent organic pollutants. Between 1992 and 2007 the occurrence of ice break-up went from mid-July to mid-June. This difference resulted in polar bears hunting more for water-associated seals than ice-related species, and open water prey have higher concentrations of PBDEs and chlorinated pesticides.

Often complex relationships exist between contaminants and ecosystem dynamics. Walters et al. (2015) conducted an experiment to evaluate primary productivity effects on methylmercury (MeHg) accumulation in stream invertebrates. They set up different lighting conditions to produce a gradient in primary productivity as measured by oxygen release. Three two-level food webs were established consisting of phytoplankton/filter feeding clams, periphyton/grazing snail, and leaf/shredding amphipod (*Hyalella azteca*). Periphyton are algae that are attached to surfaces such as sediment floors or aquatic vegetation. Phytoplankton are in the water column. Phytoplankton and periphyton biomass, along with MeHg removal from the water column (by the algae or consumers), increased significantly with productivity, but MeHg concentrations in these primary producers declined. MeHg concentrations in clams and snails also declined with productivity, and consumer concentrations were strongly correlated with MeHg concentrations in primary producers. Consumer biomass on leaves, MeHg in leaves, and MeHg in *Hyalella* were unrelated to stream productivity. The results supported the hypothesis that contaminant bioaccumulation declines with stream primary production because the contaminant load is shared among an increasing biomass of algae (MeHg burden per cell decreases in algal blooms), and this reflects what has been found in lakes and ponds.

The bottom line to the discussion on higher level effects of contaminants to populations, communities, and ecosystems goes back to the introduction of this chapter. All levels of scale from the subcellular to the ecosystem are intrinsically related. Changes due to contaminants at the individual level may ultimately alter conditions in ecosystems.

10.4   CHAPTER SUMMARY

1. All ecological components are hierarchical and related. Negative effects produced by contaminants at the cellular or subcellular level may eventually impact entire organisms. Death or disease in individuals may accumulate to negatively affect entire populations. As populations suffer, communities may be disrupted, and as entire communities are altered, significant changes may occur at the ecosystem level.
2. Contaminants can exert multiple effects on all aspects of communities. The most obvious harm is extinction or local extirpation of entire populations. Such decimation can lead to imbalances in the entire community. Contaminants can disrupt symbiotic relationships; set succession back, sometimes intentionally; and influence community dynamics in other ways. Some communities are particularly sensitive to the effects of contaminants, especially if keystone species are adversely impacted.

3. Similarly, ecosystem functioning can be harmed in many ways by high concentrations of contaminants. Because contaminants are chemicals or minerals (e.g., metals), they rightly belong under the umbrella of ecosystem ecology. Specific aspects of ecosystems that can be sensitive to pollution are energy transfers, productivity, respiration, and the efficiency of nutrient and mineral cycling.

4. Global climate change is an ecosystem problem brought on by greenhouse gases, a set of contaminants that include carbon dioxide, methane, water vapor, and nitrous oxide. While global climate change is due to a combination of natural and anthropogenic factors, increasing evidence suggests that much, perhaps most, of the problem is due to human-caused pollution stemming from the combustion of fossil fuels.

10.5   SELF-TEST

1. Describe in your own words how cellular physiology and ecosystem effects are connected in an ecotoxicological context.
2. The occurrence of food webs in most communities can
   a. Lead to the distribution of contaminants in many different species
   b. Allow for a buffering effect so that no one species receives a 'full load' of contaminants
   c. Have no effect on the distribution of contaminants in the biotic or living portion of an ecosystem
   d. More than one above
3. According to the study by Baguley et al. (2015), which of the following community factors was altered by oil from the BP Deepwater Horizon blowout?
   a. Nematodes became dominant species
   b. Species diversity increased in highly polluted areas
   c. Species richness increased in areas with low pollution
   d. None of the above
4. Movement and increasing concentrations of contaminants from lower to higher levels of food webs is called
   a. Bioaccumulation
   b. Assimilation
   c. Biomagnification
   d. Bioconcentration
5. Which of the following statements about arbuscular mycorrhizae (AM), corn, and ZnO nanoparticles is true?
   a. Nanoparticles were toxic to both AM and corn.
   b. Plants inoculated with AM were more resistant to ZnO nanoparticles than those without AM.
   c. The study presented data on how contaminants can affect symbiotic relationships.
   d. All of the above are true.

6. True or False. Landowners or managers frequently use herbicides intentionally to set ecological succession back to earlier stages.
7. In your own words describe the findings of Bendis and Relyea (2016) including both community and ecosystem effects.
8. What is (are) the chief causes of global climate change according to the International Panel on Climate Change?
   a. Sulfur dioxide
   b. Carbon dioxide
   c. Carbon monoxide
   d. All of the above
9. What happened to the concentration of methylmercury (MeHg) in individual organisms as primary productivity increased?
   a. The concentration tended to decline in plants and animals due to increased numbers of individuals.
   b. The concentration increased in plants but declined in herbivores.
   c. Concentrations of MeHg increased in all segments of the ecosystem.
   d. Concentrations declined in animals but not plants.
10. True or False. Contaminants often cause changes in the P/R ratio of communities.

REFERENCES

Baguley, J.G., Montagna, P.A., Cooksey, C., Hyland, J.L., Bang, H.W. et al. 2015. Community response of deep-sea soft-sediment metazoan meiofauna to the Deepwater Horizon blowout and oil spill. *Marine Ecol Prog Ser* 5298: 127–140.

Bendis, R.J., Relyea, R.A. 2016. If you see one, have you seen them all?: Community-wide effects of insecticide cross-resistance in zooplankton populations near and far from agriculture. *Environ Poll* 215: 234–246.

Borja, A., Dauer, D.M., Elliott, M., Simenstad, C.A. 2010. Medium- and long-term recovery of estuarine and coastal ecosystems: Patterns, rates and restoration effectiveness. *Estuaries Coasts* 6: 1249–1260.

Ebert, D. 2006. Daphnia magna asexual.jpg. Basel, Switzerland. https://commons.wikimedia.org/w/index.php?curid=47132022.

Gross, L. 2007. Diego Fontaneto—Who needs sex (or males) anyway? *PLoS Biol* 5(4): e99 doi:10.1371/journal.pbio.0050099.

Gustafson, K.D., Belden, J.B., Bolek, M.G. 2016. Atrazine reduces the transmission of an amphibian trematode by altering snail and ostracod host-parasite interactions. *Parasitol Res* 115: 1583–1594.

Intergovernmental Panel on Climate Change (IPCC). 2014. Climate change 2014: Synthesis report. Contribution of Working Groups I, II and III to the Fifth Assessment Report of the Intergovernmental Panel on Climate Change (Core Writing Team, Pachauri, R.K. and Meyer, L.A.). Intergovernmental Panel on Climate Change, Geneva, Switzerland, 151pp.

Johnston, E.L., Mayer-Pinto, M., Crowe, T.P. 2015. Chemical contaminant effects on marine ecosystem functioning. *J Appl Ecol* 52: 140–149.

Lin, Q.X., Mendelssohn, I.A., Graham, S.A., Hou, A.X., Fleeger, J.W., Deis, D.R. 2016. Response of salt marshes to oiling from the Deepwater Horizon spill: Implications for plant growth, soil surface-erosion, and shoreline stability. *Sci Tot Environ* 557: 369–377.

Mckinney, M.A., Peacock, E., Letcher, R.J. 2009. Sea ice-associated diet change increases the levels of chlorinated and brominated contaminants in polar bears. *Environ Sci Technol* 43: 4334–4339.

Olson, P.E., Fletcher, J.S. 2000. Ecological recovery of vegetation at a former industrial sludge basin and its implications to phytoremediation. *Environ Sci Pollut Res* 7: 195–204.

Perry, M.C. 2007. Changes in food and habitats of waterbirds, Chapter 14. Pp. 60–62. In: Phillips, S.W. (ed.), *Synthesis of U.S. Geological Survey Science for the Chesapeake Bay Ecosystem and Implications for Environmental Management*. U.S. Geological Survey Circular 1316, 63pp.

Sparling, D.W. 2016. *Ecotoxicology Essentials*. Boston, MA: Academic Press.

Walters, D.M., Rosi-Marshall, E., Kennedy, T.A., Cross, W.F., Baxter, C.V. 2015. Mercury and selenium accumulation in the Colorado River food web, Grand Canyon, USA. *Environ Toxicol Chem* 34: 2385–2394.

Wang, F.Y., Liu, X.Q., Shi, Z.Y., Tong, R.J., Adams, C.A., Shi, X.J. 2016. Arbuscular mycorrhizae alleviate negative effects of zinc oxide nanoparticle and zinc accumulation in maize plants—A soil microcosm experiment. *Chemosphere* 147: 88–97.

Zheng, B.H., Zhao, X.R., Ni, X.J., Ben, Y.J., Guo, R., An, L.H. 2016. Bioaccumulation characteristics of polybrominated diphenyl ethers in the marine food web of Bohai Bay. *Chemosphere* 150: 424–430.

11 Risk Assessment

11.1 INTRODUCTION

In addition to those scientists who study the fate and transport of chemical contaminants in the environment or who determine the effects of exposure of these chemicals to target organisms, there is another branch of ecotoxicology usually called risk assessment or risk analysis. Practitioners of this discipline may ask questions such as the following:

1. What is the probability that certain species/communities/ecosystems will be negatively impacted by the presence of this landfill if it suffers a breach?
2. What are the potential hazards of building a paper mill on this river?
3. What will the cost of remediating this site be and how does that figure into our overall calculations for making a profit?
4. What steps can we take to reduce or even eliminate harm from a particular hazard?
5. What are the likely ecological or human effects from this dam breaking and releasing hundreds of thousands of gallons of mining sludge?

A working definition of **ecological risk assessment** is "the process for evaluating how likely it is that the environment may be impacted as a result of exposure to one or more environmental stressors such as chemicals, land change, disease, invasive species and climate change" (EPA, 2016a). In the broad sense, these stressors include weather, natural disasters, contaminants, and a host of other factors. There is also human risk assessment, but in this chapter, of course, we are mostly interested in hazards posed by chemical contaminants to the environment, but the principles apply to most forms of risk assessment. Risk assessments in this context may cover potential harm from existing conditions such as a landfill or an oil spill into an estuary. They may also evaluate potential risks from the construction of new facilities. Assessments also determine the best management practices for preventing pollution. And they may evaluate the costs and feasibility of cleaning up an already contaminated site as well as other scenarios, but this gets us into a different phase of the overall process called risk management (see below).

There are a couple of related terms that have been tossed around that pertain to risk assessment. '**Risk**' is frequently defined as the probability of occurrence of an adverse effect on the environment as a result of a hazard. In general, a '**Hazard**' is defined as a condition, process, or state that adversely affects the environment. To clarify any circular thinking, hazard is the thing in the environment that may cause harm, and risk is the probability of its occurring. Environmental hazards may take many forms, ranging from habitat destruction, shoreline erosion, poor farming practices, wildfires, and other factors that may disrupt environmental conditions involving fish and wildlife. In this chapter we will pretty much stick with chemical hazards.

Three major considerations go into chemical risk assessments (EPA, 2016a): (1) the concentration of a chemical present in an environmental medium (e.g., soil, water, air), (2) how much exposure an ecological receptor (organism, etc.) has with the contaminated environmental medium, and (3) the inherent toxicity of the chemical. Risk is reduced by low concentrations of low toxicity chemicals that come in limited contact with receptors; it is increased by highly toxic chemicals present freely in the environment at high concentrations.

11.2 BRIEF OVERVIEW OF THE HISTORY OF RISK ASSESSMENT

Rigorous environmental risk assessment of chemical hazards is a comparatively recent outgrowth of ecotoxicology. Whereas risk assessment has been around for several decades, early studies focused on human risk assessment in medical technology or workplace environments. Arguably, an early awareness of environmental risk assessment came with a 1972 amendment to the Federal Insecticide, Fungicide, and Rodenticide Act (FIFRA), first passed in 1947. The amendment, otherwise known as the Federal Environmental Pesticide Control Act (1972), gave the Environmental Protection Agency (EPA) the job of considering the risks and benefits of pesticides when imposing regulations. Early attempts at risk assessment lacked strong guidance and were a hodgepodge of methods to evaluate perils to the environment caused by contaminants.

The National Research Council (NRC, 1983) published one of the first formal recommendations for conducting formal risk assessments. That publication was a starting point that motivated numerous investigations dealing with risks associated with chemical stressors. Glenn W. Suter and his colleagues published several papers and seminal books (e.g., Suter et al., 1987, 2000; Suter, 2006) that further laid the foundation for risk assessment. The U.S. EPA also developed several publications further outlining how to proceed with assessments (U.S. EPA, 1986, 2015, and others). Today many risk assessments fall into a regulatory context where perpetrators of contaminated environments are legally required to remediate their pollution if they are found culpable. Such remediation activities may cost millions, even billions of dollars, so rigid protocols have to be followed to make sure that they can stand against rigorous examination. According to an Associated Press article, for example, in 2014, BP Oil declared that the total outlay for damages due to the Deepwater Horizon oil spill would be $61.6 billion.

11.3 WHAT IS AN ECOLOGICAL RISK ASSESSMENT LIKE?

Risk assessments have to be tailored to the conditions at hand. Different procedures will be used for aquatic environments than for terrestrial sites or atmospheric contamination. The type of chemicals, actual or suspected, will also determine how the process will occur. In addition to general ecological conditions, risk assessment processes must also consider whether species of concern (e.g., threatened or endangered species) are present and what special accommodations are necessary for them.

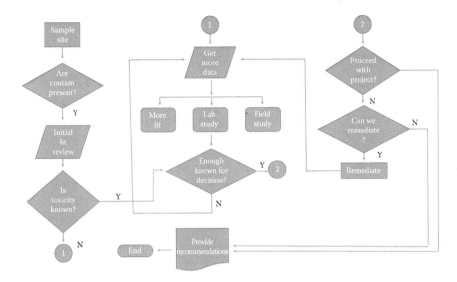

FIGURE 11.1   A simplified flow chart for performing risk assessments. See text for explanation.

There really are so many considerations that we can only provide a very general or the risk assessment process.

A very simplified risk assessment process might follow something like that shown in Figure 11.1. A primary action would be to sample the site, looking for and quantifying suspected contaminants based on preliminary information gathered about the area. For projects that may produce contamination, the concentration of added pollution needs to be factored into existing levels. Because analyses can be very expensive, one might want to avoid sampling for the several thousand contaminants that are produced each year and home in on the most likely groups of chemicals. If the site is downstream from a mine, metals would be an obvious choice of possibilities. If it were located near a plant that produced pressure-treated lumber, arsenic and toxic polycyclic aromatic hydrocarbons might first get attention. And if it were in an agricultural location, pesticides could be suspected.

If contaminants are detected, a likely follow-up question based on preliminary literature reviews and firsthand knowledge would be whether their concentrations are sufficiently high to cause concern. Perhaps the toxicity of the existing contaminants to the organisms that live in the habitat is not sufficiently known to make a decision about proceeding further. In that case, the investigators may want to gather more information. This might involve doing a more thorough literature review, or performing appropriate laboratory or field studies, or a combination of all three. A good literature review could suggest what endpoints to look for. As we have discussed, lethal toxicity might occur at very high, environmentally unrealistic concentrations, but sublethal effects such as endocrine disruption or tumor production might occur at the concentrations of contaminants in the habitat. After the experiments and literature reviews, it is possible that sufficient information is still unknown and further refined studies using larger sample sizes may be required.

If there is sufficient information on the toxicity of the existing or projected concentrations of contaminants, the risk assessment can proceed to the next stage—decision making. This actually falls into the category of risk management, but we include some discussion here to complete the description of the process. Perhaps the levels of contaminants are high enough to be toxic to organisms living in the habitat. If so, a cost-benefit analysis might have to be conducted (not shown). The benefits of conducting the project may outweigh the costs to the ecological receptors that may be affected. For instance, if a landfill for a large city is being planned, the benefits to the inhabitants and businesses of that city may be much greater than the costs to the invertebrates that would be impacted. Or, after careful consideration, project managers may decide that an area cannot be remediated or that the costs simply outweigh any potential benefits, and they scrap the project or find a different location. In such cost-benefit analyses, species of concern may be of paramount consideration and the project would have to be stopped. Alternatively, when potential benefits are great and there is risk to species of concern, project managers may evaluate the possibility of remediating the site, that is, cleaning it up. This can be a very expensive undertaking, so cost is often a factor in these decisions. Perhaps the project can be moved to another site where the risk to species or the possible contamination would be less.

An important step not included in Figure 11.1 and too often overlooked is continued monitoring after the assessment is completed. Various problems can occur, such as a breach in a landfill or containment dam, new sources of pollution can arise, or the chosen method of preventing further contamination may have been inadequate.

As we mentioned, risk assessments and associated activities can be very expensive. Even a moderate project can cost a million dollars or more. Needless to say, these costs are constantly of concern to project managers and play important roles in determining the shape of risk assessments and remediation; conscientious companies want to get the process done correctly the first time.

11.4   ASSESSING RISK TO ORGANISMS

Once the chemicals in a polluted environment have been identified and their concentrations in various environmental media (soil, water, air) have been measured, risk assessments most often become concerned with estimating the harm that organisms, such as fish or wildlife, encounter by the presence of these chemicals. In this section, we discuss how that can be accomplished. For professionals, there are mathematics and modeling involved, which some students will handle easily, but for those that find math difficult, we have tried to minimize the amount of mathematics as much as possible.

An early step in determining risk to organisms is to estimate their amount of exposure, or dose if you like, that they are receiving. From a mathematical approach this can be expressed in the form of (Equation 11.1)

$$E = \frac{1}{T}\sum C_{ijk} t_k \qquad (11.1)$$

where
    E is the exposure concentration or exposed dose
    T is the total time and space over which the concentrations in various microenvi-
        ronments or habitats are to be averaged
    C_{ijk} is the concentration in microenvironment k that is linked to environmental
        matrix i by pathway j
    t_k is the time and space that accounts for a receptor's contact with specific micro-
        environment or habitat k

This equation can be revised to one that is more explanatory than mathematical
(Equation 11.2):

$$ED = Der_{ed} + Inh_{ed} + f(Inh_{ed}) + Ing_{ed} + f(Ing_{ed}) + DW_{ed} + f(DW_{ed}) \quad (11.2)$$

where
    ED is the exposed dose
    Der_{ed} is the dermal or cutaneous exposed dose
    Inh_{ed} is the inhalation exposed dose
    $f(Inh_{ed})$ is the exposed dose coincidental to inhalation
    Ing_{ed} is the ingestion exposed dose
    $f(Ing_{ed})$ is the exposed dose coincidental to ingestion
    DW_{ed} is the drinking water consumption exposed dose
    $f(DW_{ed})$ is the exposed dose coincidental to drinking water consumption

We can get a graphic representation of this model by using its principle elements
(Figure 11.2). Once the input parameters are quantified, the risk assessor can derive
an exposure index or exposed dose. To even more fully understand the dynamics of
this exposure, however, we could add in the processes that lead to the elimination or

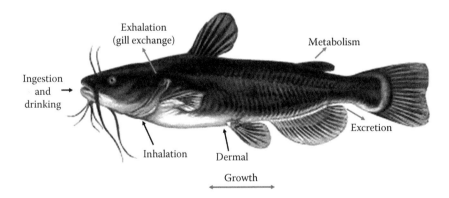

FIGURE 11.2   An organism-based risk assessment should consider all of the various entry
and exit areas for the contaminant of interest and develop a mass balance for the chemical and
organism model.

dilution of the contaminants, which include exhalation (gill exchange in the case of fish), metabolism of the contaminants, excretion, and growth (which can cause dilution if the rate of growth exceeds the rate of intake). If rates can be applied to each of these input and output factors, a **chemical balance equation** can be developed, which can be used to predict the rate of transport of contaminants in the organism. Such equations can only be developed with any sense of accuracy in the laboratory and then roughly transferred to field situations. Note that total chemical balance equations, while an outgrowth of the risk assessment process, are usually not included in the assessments in the attempt to keep costs down and regulatory agencies as a rule do not require them.

Once an estimate of exposure has been calculated, we need an estimate of overall toxicity or, in the parlance of risk assessment, a **toxicity reference value** (TRV). In many cases, these TRVs might not be derived from the actual species of concern in a given assessment. Rather, they may be taken from one of several species that the U.S. EPA normally uses as reference animals. These include certain invertebrates, fish, birds, and mammals that are easily maintained under laboratory conditions and for which a wealth of toxicity data is known. In a regulatory process, these model species have to bear some similarity to the species of concern, a risk assessor cannot use a mallard duck in place of fish, for example, but they do add some measure of unknown since toxicity values can differ by orders of magnitude even within the same class of organisms. These TRVs can provide a generalized understanding of how exposure to selected chemicals in the environment translates into risks.

A commonly applied approach for evaluating ecological risks from environmental chemicals relies on a simple ratio (U.S. EPA, 2015) that estimates the potential for harmful effects to an organism by comparing the organism's estimated levels of exposure to appropriate TRVs. Most TRVs for chemicals commonly encountered in environmental settings of regulatory interest have been derived from published toxicity studies, and a representative value is selected as a benchmark TRV for regulatory applications. These data are used to calculate a **hazard quotient** (HQ) as follows (Equation 11.3):

$$HQ = \frac{\text{Site exposure level}}{\text{TRV}} \tag{11.3}$$

If the value of HQ is less than or equal to 1.0, risk is considered negligible to low and that no unacceptable impacts should occur in the exposed population of organisms. In contrast, if the value of the HQ exceeds 1.0, then an unacceptable impact may occur, and risk will be moderate to high. As suggested above, caution must be used with such broad-brush approaches. In addition to the species differences, TRVs should be determined from organisms of similar life history stages (e.g., young organisms tend to be more sensitive than adults and there may be differences in toxicity due to sex or other factors). Also, animals with life history attributes that markedly differ from the chosen few may not fit well with the hazard quotient concept. For example, TRVs for dermal exposures derived from fish may not be appropriate for amphibians that have a more permeable skin and rely extensively on dermal respiration (Linder et al., 2010). A huge potential problem with the TRV process is that it has evolved for only one contaminant being present, but in reality most contaminated sites are going to have multiple chemicals present with all sorts of interactions.

11.5   UNCERTAINTY IN RISK ASSESSMENTS

We would seriously mislead you if it seems that risk assessment is cut and dried. Risk assessment is not at the stage that assessors can simply plug values into a computer program and have it print out an acceptable level of contamination in every environmental context. Many decisions are still dependent on the 'art' of risk assessment, meaning that assessors sometimes have to use their Best Professional Judgment (aka 'wild ass guess'), especially in the preliminary phases of an assessment. However, uncertainty continues to pop up throughout the process. Subsequently, it may behoove us to spend a little space discussing uncertainty as it applies to this field of science.

Two general types of uncertainty exist—aleatory and epistemic. **Epistemic uncertainty** is due to a lack of sufficient information. This may be due to inadequate sample sizes or duration of preliminary studies or a lack of previously published studies on problems similar to those at hand. Also, as we have mentioned, organisms are seldom exposed to only one contaminant at a time. For example, plants growing near a landfill might be exposed to lead, mercury, other metals, and several organic contaminants. While there might be relevant information on the responses of the plants to some of these by themselves, little to no data may be available on the various interactions that can occur among the contaminants present. Perhaps the assessors are working with relatively new chemicals for which little data can be found. When possible, regulatory agencies can try to reduce this type of uncertainty by requiring adequate sample sizes and appropriate study durations, but some of the relevant information is just not available without prohibitory cost.

Aleatory uncertainty is due to the inherent randomness of a system. In an environmental context, this includes all of the seemingly random and not-so-random factors that can impinge on an environmental receptor. If an assessment is concerned about whether the population of a keystone species is changing, it would have to acknowledge effects due to weather, predation, food bases, and so on. It may sound preposterous, but an earthquake could cause upheaval several miles upstream that results in a flush of new contaminants into the study area—this could really mess up the understanding of post-remediation processes.

Characterizing uncertainty in many decision-making efforts is seldom fully appreciated. Attempts to make valid decisions can be seriously hampered if uncertainty is not at least partially recognized. Some types of uncertainty are more easily controlled than others. For example, making sure that protocols are carefully followed, that good laboratory practices (see OECD, 2015) are adhered, that measurements are precise, and that sample sizes are adequate can reduce a considerable amount of uncertainty. In contrast, epistemic uncertainty arises through lack of knowledge or scientific ignorance and reflects uncertainty related to state-of-knowledge rather than the lack of knowledge. Therefore, epistemic uncertainty is frequently termed irreducible uncertainty. Using the examples above, we might not have any idea of how many samples will be needed or what level of accuracy in measurements is required before we go into a study. There are ways of evaluating the quality of data and initial efforts may indicate the need to conduct other studies, but this is the job of the risk manager (see below) and he/she has to make decisions based in part on costs. Maybe the manager will decide that the data are not as good as they could be but are adequate. Maybe he/she will be wrong; it has happened more than once.

11.6 VULNERABILITY ANALYSIS

Not all species are created equal, nor are all habitats. Some species may be very sensitive to a given contaminant, others more tolerant. Some habitats are resilient, take a beating from contaminants or other stressors and bounce back, other environments may be less likely to rebound. Changing conditions such as the addition of other stressors can increase the sensitivity of a system to pollution. **Vulnerability analysis** (Figure 11.3) attempts to account for this variability in ecosystem elements and adjust risk assessments accordingly. Vulnerability is registered not by exposure to hazards (perturbations and stresses) alone but also by the sensitivity and resilience of the system experiencing such hazards (Turner et al., 2013).

A couple of analogies may help understand this concept of vulnerability. We can simplify the response of a species or any environmental element to disturbance in two ways—resistance and resiliency. Let's collectively call the species or environmental element the receptor. **Resistance** is measured by a receptor's ease of changing. In terms of contaminants this may also be called sensitivity. A receptor may

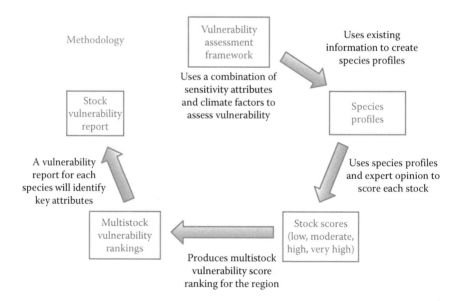

FIGURE 11.3 A vulnerability assessment for oceanic fish species exposed to climate change. Each species is evaluated separately but combined assessments are used for multiple species fish stocks. (Courtesy of NOAA, *Assessing the Vulnerability of Fish and Invertebrate Species in a Changing Climate*, N.D., https://www.st.nmfs.noaa.gov/ecosystems/climate/activities/assessing-vulnerability-of-fish-stocks.)

have high resistance or low sensitivity to a contaminant and will not respond until concentrations become very high, perhaps higher than what is environmentally realistic. In contrast, receptors with low resistance will react negatively even at low concentrations. While it may be easy to see this relationship in organisms, habitats also display varying degrees of resistance. Wetlands are sensitive to a wide variety of stressors, including contaminants. Deserts, on the other hand, are quite resistant to most stressors.

The other element is **resiliency**, which is the ability of a receptor to rebound after a disturbance has occurred. Species with high fecundity rates generally have a much greater ability to rebound after some disturbance than species with low reproductive output. However, long-lasting contaminants may produce multigenerational problems. Some habitats can similarly rebound much quicker than others; grasslands have a greater resiliency to any decimating stressor than mature forests, for example.

11.7   RISK MANAGEMENT

Once an analysis of exposures and effects is completed and risk has been determined, **risk management** takes the lead. Risk assessment and management are two sides of the same coin (Figure 11.4). Once risk has been determined, affected receptors have been identified, and the chemicals associated with that risk have been determined and quantified, risk management kicks in. On paper, risk management seems to be a relatively straightforward process that focuses on protecting human and environmental health (Figure 11.5). As with most things, however, the actual process can be much more complicated and entails financial and legal components. For example, risk management may have to follow governmental regulations on what wastewater discharge limits are acceptable for mining activities and then determine the necessary costs to reduce or maintain the discharges at acceptable levels. Alternatively, the EPA

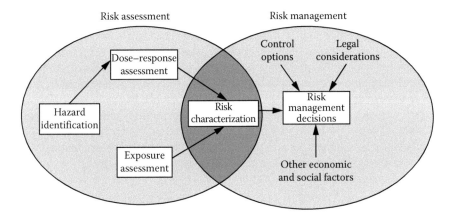

FIGURE 11.4   A model showing the major components of risk assessment and management and how the two disciplines interact. (Courtesy of EPA, National Resource Council (NRC), *Risk Assessment in the Federal Government: Managing the Process,* National Academic Press, Washington, DC, 1983, https://www.epa.gov/fera/nrc-risk-assessment-paradigm.)

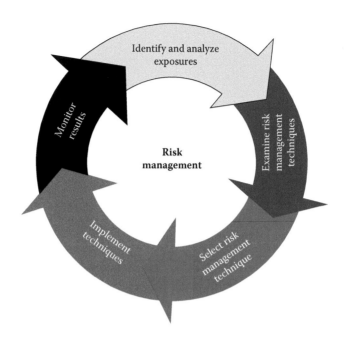

FIGURE 11.5 A schematic showing the process of risk management. After the contaminants, concentrations, and exposures have been identified, the risk manager begins to assess the risk to the environment. Specific methods are selected and implemented to remediate these risks. Monitoring the end result is very important to assure that the remediation has been effective. (Courtesy of Bmet Wiki, Risk Management, 2010, http://bmet.wikia.com/wiki/Risk_Management.)

or other agency may mandate that a company repair a leaking landfill, and the risk manager will have to propose an economically viable way of doing that.

A part of the risk management process is evaluating **acceptable risk limits**. It is very unlikely that any contaminated environment can be totally remediated. Environments are complex, and for example, heavy metals that were bound in the sediment of a river at the time of the assessment might be recycled due to future disturbances that are either natural or human caused. Thus, risk managers often decide on a level of risk that is acceptable; this level commonly has to be approved by a regulatory agency. These acceptable levels are based on the type of risk since many different risks may potentially be applicable. Depending on the nature of the environment, the contaminants that are present, and the species of concern that are there, acceptable levels of risk may be less than 1% or as high as, say, 10%.

The risk management process will determine what to do after the assessment has been made. Perhaps the area in question is under regulatory constraints and the polluter has been identified. In these cases, the offending party may be required to clean up the area to specified standards. In other cases, a specific party or group of parties have not been identified, but the area is still under regulatory authority, in which case cleanup falls under the jurisdiction of the government and the site is placed in priority with thousands of other contaminated sites. If the area is relatively small and not connected in any way (including possible leakage into streams) to any interstate water course, it may not be under federal regulatory authority and cleanup may fall under varying state laws or to the ethical nature of the landowner.

11.8   CHAPTER SUMMARY

1. Risk assessment is the application of scientific principles to determine the chance of harm to human or ecological health.
2. Early risk assessments were conducted without much formality. Today, rigorous protocols developed by the U.S. EPA and other organizations take the practitioner through each step. Yet, assessments still involve some 'art' in deciding which parameters to measure and what to do when encountering a situation that lacks adequate information.
3. Risk assessments take many forms and involve all sorts of scenarios. Those concerned with chemical hazards are only one category.
4. An essential part of risk assessment is to determine the risk that ecological receptors such as organisms or habitats encounter in a contaminated environment. This is often accomplished by using models and mathematics. Toxicity reference values are frequently employed, which use estimates of toxicity developed from previously published data, and these along with data on contaminant concentrations are used to developed hazard quotients.
5. Uncertainty is always present in scientific studies including risk assessments. Aleatory uncertainty occurs due to random processes in any environmental study, and epistemic uncertainty happens when there is insufficient data. Risk managers must cope with both types of uncertainty, but established protocols help reduce the amount of uncertainty in such appraisals.
6. Increasingly, vulnerability analysis is becoming a component of risk assessments when risk managers and risk assessors account for variability among organisms and other ecological receptors to the presence of contaminants.
7. Risk management takes over once assessment has identified real or potential harm to ecological receptors and characterized the stressor (chemicals in our case) and their strengths or concentrations. Risk managers must take into account the best methodologies to remediate a problem and consider legal and financial factors during the entire process.

11.9   SELF-TEST

1. What is the difference between a risk and a hazard?
   a. Risk is the probability that some harm may occur; hazards are the cause of that harm.
   b. The two terms are correctly used interchangeably in risk assessments.
   c. Hazards are natural physical objects in water or land that can cause damage; risk is due to conscientious pollution.
   d. More than one above.

2. List the three elements that go into a risk assessment as defined by the U.S. EPA.

3. What does the acronym FIFRA stand for?
   a. Federal Information on Fungus, Rodents, and Agriculture
   b. Federal Insecticide, Fungicide, and Rodenticide Act
   c. Fungicide, Insecticide, Foliar, and Rodenticide Act
   d. None of the above

4. A first step in any risk assessment investigation should be to
   a. Study the published literature on similar situations
   b. Make a thorough evaluation of the contaminants at the site
   c. Conduct laboratory studies to see how sensitive organisms that inhabit a site are to several contaminants
   d. Decide if the site can be remediated

5. True or False. Costs seldom enter into making decisions about what to do with a contaminated site costs.

6. In the equation $E = \dfrac{1}{T} \sum C_{ijk} t_k$, E stands for
   a. Entropy
   b. Enhancement potential
   c. Exposure
   d. Extinction probability

7. Short answer. What is a toxicity reference value (TRV) and what is it used for?

8. At what point is the value of a hazard quotient of concern?
   a. When it is below 0
   b. When it is below 1
   c. When it exceeds 1
   d. Not until it exceeds 10

9. Which of the following would add to epistemic uncertainty?
   a. A forest fire occurs, adding polycyclic aromatic hydrocarbons to a study site.
   b. A study design has a sample size that is too small to adequately describe the site.
   c. A new chemical is found whose effects are not well known.
   d. b and c above
   e. All of the above

10. In a risk assessment that you have been commissioned to do, you acknowledge that some of the species at the site have very low tolerance to the chemicals that are present, whereas others have substantially higher tolerances. This distinction would be a part of
    a. Toxicity reference values
    b. Hazard analysis
    c. Vulnerability analysis
    d. Tolerance analysis

REFERENCES

EPA, National Resource Council (NRC). 1983. *Risk Assessment in the Federal Government: Managing the Process.* Washington, DC: National Academic Press. https://www.epa.gov/fera/nrc-risk-assessment-paradigm.

Federal Environmental Pesticide Control Act. 1972. (7 USC 136-136y, P.L. 92-516, October 21, 1972, 86 Stat. 973) as amended by: P.L. 93-205, December 28, 1973, 87 Stat. 903; P.L. 94-140, November 28, 1975, 89 Stat. 751; P.L. 95-396, September 30, 1978, 92 Stat. 819; P.L. 98-201, December 2, 1983, 97 Stat. 1379; and P.L. 100-202, December 22, 1987, 101 Stat. 1329.

Linder, G., Palmer, B.D., Little, E.E., Rowe, C.L., Henry, P.F.P. 2010. Physiological ecology of amphibians and reptiles: Life history attributes framing chemical exposure in the field. Pp. 105–166. In: Sparling, D.W., Linder, G., Bishop, C.A., Krest, S.K. (eds.), *Ecotoxicology of Amphibians and Reptiles*, 2nd Edition. Boca Raton, FL: CRC Press.

NOAA. N.D. *Assessing the Vulnerability of Fish and Invertebrate Species in a Changing Climate.* https://www.st.nmfs.noaa.gov/ecosystems/climate/activities/assessing-vulnerability-of-fish-stocks.

Organisation for Economic Co-operation and Development (OECD). 2015. Good laboratory practice (GLP). http://www.oecd.org/chemicalsafety/testing/goodlaboratorypracticeglp.htm. Accessed November 26, 2016.

Suter II, G.W. 2006. *Ecological Risk Assessment*, 2nd Edition. Boca Raton, FL: CRC Press, 680pp.

Suter, G.W., Barnthouse, L.W., O'Neill, R.V. 1987. Treatment of risk in environmental-impact assessment. *Environ Manage* 3: 295–303.

Suter II, G.W., Efroymson, R.A., Sample, B.E., Jones, D.S. 2000. *Ecological Risk Assessment for Contaminated Sites*. Boca Raton, FL: CRC Press, 460pp.

Turner II, B.L., Kasperson, R.E., Matson, P.A., McCarthy, J.J., Corell, R.R. et al. 2013. A framework for vulnerability analysis in sustainability sciences. *Proc Natl Acad Sci* 100: 8074–8079.

U.S. Environmental Protection Agency (EPA). 1986. Hazard evaluation division. Standard evaluation procedure, ecological risk assessment. EPA 540/9-85-001. Prepared by Urban, D.J., Cook, N.J. Washington, DC: Office of Pesticide Programs. U.S. Environmental Protection Agency, 103pp. Accessed November 26, 2016. https://nepis.epa.gov/Exe/ZyNET.exe/91012PC3.TXT?ZyActionD=ZyDocument&Client=EPA&Index=1986+Thru+1990&Docs=&Query=&Time=&EndTime=&SearchMethod=1&TocRestrict=n&Toc=&TocEntry=&QField=&QFieldYear=&QFieldMonth=&QFieldDay=&IntQFieldOp=0&ExtQFieldOp=0&XmlQuery=&File=D%3A%5Czyfiles%5CIndex%20Data%5C86thru90%5CTxt%5C00000026%5C91012PC3.txt&User=ANONYMOUS&Password=anonymous&SortMethod=h%7C-&MaximumDocuments=1&FuzzyDegree=0&ImageQuality=r75g8/r75g8/x150y150g16/i425&Display=hpfr&DefSeekPage=x&SearchBack=ZyActionL&Back=ZyActionS&BackDesc=Results%20page&MaximumPages=1&ZyEntry=1&SeekPage=x&ZyPURL.

U.S. Environmental Protection Agency (EPA). 2015. *Risk Characterization Handbook.* https://www.epa.gov/risk/risk-characterization-handbook. Accessed November 26, 2016.

U.S. Environmental Protection Agency (EPA). 2016a. About risk assessment. https://www.epa.gov/risk/about-risk-assessment#whatisrisk. Accessed November 26, 2016.

12 Domestic and Global Regulation of Environmentally Important Chemicals

12.1 INTRODUCTION

Contaminants are released into the environment throughout the world, and those that enter the atmosphere or water sources have the potential to broadly spread across national borders. Consequently, agencies have been established at the international, national, and regional levels to regulate, attempt to control, and remediate these contaminants. We have discussed many times in this book the role of the U.S. Environmental Protection Agency (EPA) as the chief watchdog for pollution within the United States. The EPA has counterparts in most of the other countries; some of these agencies are effective, some not so much, depending on the importance that governments place on pollution. For instance, China and India, two of the most populated nations in the world, have severe problems with contaminants as they improve their economic status partly through increased manufacturing. In many ways, they are repeating some of the same mistakes that the United States and other developed nations did during the Industrial Revolution. Manufacturing processes almost by definition produce toxic chemicals, and the disposition of these chemicals determines whether they will be contained, safely broken down, or released into the environment. Government agencies at various levels spend billions of dollars each year to safeguard the environment, and these agencies are the focus of this chapter. As with much of the material in this book, this is not an exhaustive discussion but a summary of activities.

12.2 INTERNATIONAL AUTHORITIES

12.2.1 THE UNITED NATIONS

Arguably, the United Nations Environment Programme (UNEP) is the largest and most important authority for monitoring contaminants and providing guidance on a global level. The United Nations itself is headquartered in New York City and consists of 193 member states, but the UNEP is based in Nairobi, Kenya (Figure 12.1). Its mission statement reads, "To provide leadership and encourage partnership in caring for the environment by inspiring, informing, and enabling nations and peoples to improve their quality of life without compromising that of future generations."

FIGURE 12.1   The United Nations Environment Programme is headquartered in Nairobi, Kenya.

The UNEP helps develop global environmental priorities, implements guidelines for sustainable development across the world, and serves as an advocate for the environment. It does not have any regulatory authority per se, meaning that it cannot order a polluting nation to stop or to assess fines, but it can exert international political pressure on countries to get on board. Among the major events sponsored by the UNEP is to call nations together to develop protocols, conventions, and accords. For example, the Stockholm Convention (2001) brought more than 90 nations together to sign an agreement to reduce the presence of persistent organic pollutants (POPs, UNEP, 2016). One concern of these agreements is that member nations may sign on but then they must have their governments ratify or vote to accept the conditions of the agreement. Thus, the United States never ratified the Stockholm Convention (U.S. EPA, 2016a) nor has it ratified the Kyoto Protocol (1997).

Disputes between countries concerning these agreements can be brought before the International Court of Justice, otherwise known as the International Court of Justice or World Court, the United Nation's primary judicial body. As mentioned, the World Court has no real teeth in that any member nation may refuse to follow the court's decisions.

A few major contaminant-related activities organized by the UNEP include the following:

1. Formation of the **United Nations Framework Convention on Climate Change (UNFCCC)**, an environmental treaty established in 1992 to "stabilize greenhouse gas concentrations in the atmosphere at a level that would prevent dangerous anthropogenic interference with the climate system." The Framework Convention had an objective of reducing greenhouse gas emissions in developed countries to their 1990 levels by the year 2000, but this objective was never close to realization. The treaty merely provides a framework for negotiating specific international treaties that may set binding limits

on greenhouse gases. An early outgrowth of this convention was the **Kyoto Protocol** in 1997, which established legally binding obligations for developed countries to reduce their greenhouse gas emissions. The United States withdrew from the agreement in 2001, seeking instead to follow its own timeline of greenhouse gas reductions. Canada withdrew its participation in 2012.

2. We have already discussed the **Stockholm Convention** (2001) in previous chapters, but some detail bears repeating. This is a global treaty to reduce persistent organic pollutants (POPs) in the environment. In implementing the Convention, governments agree to eliminate or reduce the release of POPs into the environment. An initial list of a 'dirty dozen' was drawn up for elimination or significant reduction. The list of regulated POPs now numbers 31, including several that are unintended by-products of production (UNEP, 2016).

3. **The Vienna Convention** (1984) and the **Montreal Protocol** (1987) were established to control the release of greenhouse gases and other aerosols that reduce the ozone layer protecting the earth. Chief categories of concern were chlorofluorocarbons (CFCs), methane, halons, carbon tetrachloride, methane, and nitrous oxides. Understanding climate change remains a major effort of the UNEP.

12.2.2 Organisation for Economic Co-Operation and Development

The mission of OECD is to promote policies that will improve the economic and social well-being of people around the world.

OECD (2016)

This international organization consists of 35 member nations and is headquartered in Paris, France (Figure 12.2). It provides a forum to have governments work together to share information and seek solutions to common problems. There is an emphasis on economic development but that also includes global contaminant issues. The Organisation for Economic Co-operation and Development (OECD) is not an enforcement agency, but it conducts studies on contaminants and provides guidance on contaminant issues. Among the products the OECD produces are guidelines for contaminant testing, descriptions of good laboratory practices, and issuance of legal opinions.

FIGURE 12.2   Part of OECD is headquartered in the Château de la Muette, in Paris.

FIGURE 12.3   Within the impressive building for the European Union in Belgium is its environment component, the Directorate-General for the Environment.

12.2.3   EUROPEAN UNION

The European Union (EU), founded in 1993, is a political and economic union of 28 member states (we should mention that the United Kingdom voted to leave the European Union in a vote called Brexit and that they should be completely separated from the EU in 2019) that are located primarily in Europe (Figures 12.3 and 12.4). It operates through a system of institutions and intergovernmental-negotiated decisions by the member states. The EU can fine or otherwise censor member nations that violate its directives. Within the EU is the European Commission that contains the **Directorate-General for the Environment** (European Commission, 2016). The objectives of the Directorate-General for the Environment include protecting, preserving, and improving the European environment; proposing and implementing policies to ensure environmental protection; preserving the quality of life of EU citizens; enforcing environmental regulations among member states; and representing the EU in environmental matters at international meetings. The Directorate-General for the Environment has a complex program that covers all of the major pollutants, their sources and effects.

12.3   NATIONAL REGULATION OF CONTAMINANTS

It would be a pretty safe bet that all developed countries have their own environmental and contaminant programs, including offices or directorates dedicated exclusively to chemical pollution. In addition, many poorer but developing nations also have some agencies overseeing pollution matters. Some contaminant issues affect virtually all nations. Other issues such as oil spills mostly affect oil-producing nations, but oil-importing countries can also experience oil spills. Underdeveloped nations such as South Sudan and Somalia may be concerned about pollution but do not have the resources to do much about it, they have to rely on international assistance, especially through UNEP. Here, we will focus on the United States.

In the United States, contaminant-related issues crosscut governments at the federal, state, regional, and urban levels. Additionally, we have scores of nongovernment

FIGURE 12.4   The European Union covers much of western Europe. Recently, the United Kingdom voted to exit the EU, but full exiting will not occur until sometime in 2019.

agencies that serve as watchdogs on pollution and governments. While legislators pass laws, establish agencies, and set their missions, the actual agencies that do the leg work are generally within Executive branches, directly under the President or governors.

At the federal level, the EPA is the chief enforcement and policy maker, but it works in consort with other agencies in dealing with oil spills, federal lands, clean oceans, mine-related contaminants, effects on fisheries and wildlife, and other issues. Following is a concise narrative of federal agencies directly involved with contaminant issues.

12.3.1   STATE DEPARTMENT

The principal job of the State Department is to serve as the nation's contact with other nations. This includes developing treaties, protocols, and conventions to promote global concern for the environment. The State Department addresses global environmental chemicals through programs involving energy, agriculture, marine resources, and environmental science and technology policies to address global challenges including environmental stewardship. At present, the State is particularly concerned about the global causes and effects of climate change and assists in providing scientific and economic support in reducing their emissions.

12.3.2 DEPARTMENT OF DEFENSE

The Department of Defense manages more land in the United States than any other federal agency; millions of acres are under its jurisdiction. Each of the component departments (e.g., Army, Navy, Air Force, Marines) has its own environmental office. Of major concern are the 141 Superfund sites, 6,000 hazardous waste dumps, and 39,000 contaminated areas on its various properties. Some of these sites exist on lands that are slated to or have been downsized after decades of neglect. Highly contaminated areas such as Superfund sites prevent the Department from unloading these areas until the contamination is remediated. The EPA works closely with the Department of Defense to make sure that the sites are remediated.

12.3.3 DEPARTMENT OF AGRICULTURE

This Department has the lead federal role in assuring that the agriculture within the United States remains healthy and productive. It has several offices that work with producers to provide insurance, financial assistance, and advice. Among other functions, the Department runs the Food and Nutrition Service that funds the food stamp program and provides education on nutrition to children and their parents. The list of programs within the Department is extensive, but two agencies stand out with regard to contaminants.

12.3.3.1 U.S. Forest Service

The U.S. Forest Service (USFS) owns and supervises the 154 national forests and 20 national grasslands throughout the nation. Whereas most of the lands lie east of the Mississippi River, USFS properties are found throughout the nation. The USFS does not have any specific office dealing with contaminants per se, but the agency is concerned with clean water and clean air issues. Its policy is to reduce the application of pesticides whenever possible and to use mechanical means to maintain forests. It will use insecticides during outbreaks of harmful pests but tries to use natural means to reduce infestations such as *Bacillus thuringiensis*, pheromone-activated traps, or specific insect viruses.

12.3.3.2 Natural Resources Conservation Service

Agriculture in the United States uses millions of pounds of active ingredient pesticides each year. This includes the whole list of pesticides, most notably herbicides, insecticides, and fungicides. It is critical, therefore, to have specialists who can advise growers when and what pesticides to use to economically control pests. Such advising is one of the activities of chemists and biologists working for the Natural Resources Conservation Service (NRCS). Personnel in this agency work in virtually every county that has an agricultural economic base. In addition to advising on pesticides, the staff advise on erosion control that can reduce contaminant-laden effluents into natural waterways and many other activities that directly or indirectly control pollution.

In addition to working with growers, NRCS also trains, licenses, and regulates commercial pesticide applicators to make sure that pesticides are not being misapplied.

12.3.4 DEPARTMENT OF THE INTERIOR

Historically, much of the federal research on contaminant effects on fish and wildlife occurred through this Department and the U.S. Fish and Wildlife Service. Today the research is under the jurisdiction of the U.S. Geological Survey (USGS), and the work continues, although somewhat abated.

12.3.4.1 Fish and Wildlife Service

This agency has charge of specific migratory fish and wildlife resources in the country including endangered and threatened species. It often works closely with the EPA and the National Marine Fisheries Service (NMFS) on the **Natural Resource Damage Assessment (NRDA)** issues. NRDA steps in after large-scale environmental crises such as the BP oil spill, a blowout of a mining retention pond, or some other releases of hazardous substances to calculate the monetary cost of restoring natural resources, including wildlife that were damaged by these releases. An objective of the NRDA is to try and identify and recover the costs from potentially responsible parties so that the populations and their habitats can be restored. Damages to natural resources are evaluated by identifying the functions or 'services' provided by the resources, assessing the current value of these services, and quantifying any future reductions in service levels due to the contamination. The NMFS becomes involved when marine oil spills occur, and both agencies have their specific roles.

Another important function of the U.S. Fish and Wildlife Service with regard to contaminants exists within the **Ecological Services Branch** of the agency. Each state has an Ecological Services office with staff specialists in wetlands, water resources, endangered species, conservation planning assistance, natural resource damage assessment, and contaminants. The offices provide scientific assistance to wildlife refuges, other federal and state agencies, tribes, local governments, the business community, and private citizens. Contaminant specialists in these offices review project plans including environmental assessments prior to the start of major projects and review license applications and proposed laws and regulations to avoid or minimize harmful effects on wildlife and habitats. In cases of significant releases of hazardous wastes, they often lead the efforts toward risk assessment and developing remediation plans.

12.3.4.2 Bureau of Land Management

Second only to the Department of Defense in the amount of land owned, the Bureau of Land Management (BLM) has many open pit and underground mines and oil derricks on its properties that it oversees. As a couple of examples, the Bureau has jurisdiction over 258 million acres (60 million ha) of surface lands and 700 million acres (157 million ha) of subsurface minerals primarily in the 12 western States. On these lands, there are over 46,000 oil leases and 308 coal leases totaling up to 570 million acres (106 million ha). In addition, hundreds of abandoned mines also pose risks for water pollution. The Bureau must follow federal guidelines over these activities to make sure that mining operations, landfills, surface waters, and atmospheric releases comply.

12.3.4.3   U.S. Geological Survey

According to its mission statement, the USGS serves the Nation by "providing reliable scientific information to describe and understand the Earth; minimize loss of life and property from natural disasters; manage water, biological, energy, and mineral resources; and enhance and protect our quality of life." The agency is excellent at monitoring and surveying natural resources. With contaminants, the USGS has extensive water quality monitoring programs. These programs monitor basic water quality parameters and sample for a broad range of contaminants in water and air. The USGS does not have regulatory authority, but it regularly reports its findings to agencies at the local, state, and national levels that do.

12.3.4.4   Bureau of Safety and Environmental Enforcement

Soon after the BP oil spill, the Congress approved two new agencies: the Bureau of Safety and Environmental Enforcement (BSEE) and the Bureau of Ocean Energy Management (BEOM). The BEOM was established to oversee financial aspects of outer continental shelf oil extraction and to make sure that the American people receive a fair return from leases. The BSEE is more relevant for our interests for it was set up as a watchdog on outer continental oil companies. The agency's main objectives include "(1) making sure that the energy workforce on the Outer Continental Shelf experiences the safest possible conditions; (2) protecting the environment by enforcing America's laws, driving effective prevention innovation and ensuring that companies are prepared to deal with spills, should they occur, and (3) conserving our nation's energy resources by requiring maximum feasible recovery on federal offshore lease sites and minimizing losses during the production process" (BSEE, 2016).

12.3.5   Department of Commerce

The Department of Commerce deals extensively with interstate and international trade and transport of goods. However, there are a couple of bureaus within the department that are very important in the regulating or monitoring of contaminants.

12.3.5.1   National Oceanic and Atmospheric Administration

The National Oceanic and Atmospheric Administration (NOAA) is the chief federal agency with regard to the physical aspects of oceans and the atmosphere. Whereas the Department of Interior oversees some marine activities, especially oil drilling on the continental shelf, various offices within the NOAA cover from the shoreline up to the exclusive economic zone, 200 nautical miles from shore. Contaminant-related concerns include tracking and predicting the direction of oil spills, monitoring the fate of toxic chemical spills in the ocean, helping to reduce microplastics and other debris, and monitoring the quality of tidal water. The NMFS is an agency under the NOAA and deals with the marine environment. The NOAA also tracks the effects of waterborne contaminants on marine sea life such as corals, benthic organisms, and shellfish. As part of its Atmospheric namesake, the NOAA also monitors air quality in the United

States and globally. Through their Earth Systems Research Laboratory, the NOAA regularly assesses levels of air quality elements such as tropospheric ozone, carbon monoxide, and aerosol particles across the globe. Being able to determine the status and development of global climate change is of major concern in these endeavors.

12.3.5.2 U.S. Coast Guard

An important duty of the paramilitary, U.S. Coast Guard is to assure that there are no illegal discharges of pollutants into U.S. waters. The Coast Guard is often the first federal coordinator for oil and hazardous substance spills in coastal waters. As such, the U.S. Coast Guard is the chief enforcer of an international agreement called MARPOL 73/78 or the International Convention for the Prevention of Pollution from Ships (1973, 1978). This convention was developed among 152 member nations to minimize pollution of the oceans, seas, and atmosphere. Before MARPOL, ships emptied their untreated bilge tanks directly into ocean waters under the assumption that the wastes would be diluted by the seas. They may still do this in open water where enforcement is difficult, but within the national waters of member countries, violators can face significant fines. The world has come to the recognition that dilution is not the solution for the amount of shipping that occurs.

12.3.6 U.S. Environmental Protection Agency

As the chief watchdog for contaminants, the U.S. EPA (Figure 12.5) has offices across the country to oversee remediation efforts on hazardous landfills and enforce federal regulations concerning emissions, effluents, and water treatment plants and

FIGURE 12.5 The office building for the U.S. Environmental Protection Agency, Washington, DC.

protect wetlands and virtually all things of a contaminant nature. The Congress created the U.S. EPA in 1972, and it has passed many acts that provide marching orders to the EPA in its efforts to reduce pollution and to protect human health. Below, we briefly cover a few of these acts.

12.3.6.1   Federal Food, Drug, and Cosmetic Act (FFDCA, 1938)

This is among the oldest of the acts dedicated to regulating the purity of food, cosmetics, homeopathic medicines, and bottled water. The U.S. Department of Agriculture through the Food and Drug Administration has jurisdiction over most of the Act dealing with food and some medicines. The FFDCA also authorizes the EPA to set maximum residue limits for pesticide residues in foods. These limits are adjusted as new information is acquired.

12.3.6.2   Clean Air Act (1970)

One of the first and major acts that fell under the jurisdiction of the EPA was created even before the EPA was established to monitor and regulate air quality. The Clean Air Act is one of the United States' first and most influential modern environmental laws. Major amendments to the current Clean Air Act were passed in 1977 and 1990. These amendments included coverage of acid rain, ozone depletion, standards for stationary or point sources (e.g., incinerators, electric power plants), toxic air pollutants, and increased enforcement authority. The revisions also established new auto gasoline reformulation requirements, set standards to control evaporative emissions from gasoline, and mandated new gasoline formulations sold from May to September in many states.

The Act focuses and sets standards for six major air pollutants or **criteria pollutants**—ozone, particulate matter, carbon monoxide, nitrogen oxides, sulfur dioxide, and lead. The EPA can levy fines on companies that violate the emission standards, which are updated as necessary. *De facto* additions to this list include greenhouse gases such as carbon dioxide, methane, and water vapor that are involved in global climate change, although standards have not yet been established for some of these pollutants.

12.3.6.3   Clean Water Act (1972)

Following more or less on the heels of the Clean Air Act, the Congress passed the Clean Water Act. This act monitors and regulates water pollution with objectives of (1) restoring and maintaining the chemical, physical, and biological integrity of the nation's waters by preventing point and nonpoint pollution sources; (2) providing assistance to publicly owned treatment plants for the improvement of wastewater treatment; and (3) maintaining the integrity of wetlands. It's the last of these objectives that has been the most controversial. In recent years, wetlands suitable for protection ranged from only those that were connected to rivers that could be navigated by boats to a more general description currently used by the EPA and the U.S. Army Corps of Engineers as "areas that are inundated or saturated by surface or groundwater at a frequency and duration sufficient to support, and that under normal circumstances do support, a prevalence of vegetation typically adapted for life in saturated soil conditions. Wetlands generally include swamps, marshes, bogs, and similar areas" (U.S. EPA, 2016b). The legality of what is and what is not a **protected** wetland has been the focus of these differences.

At present, the Clean Water Act consists of six programs or titles: research, grants for water treatment plants, standards and enforcement, permits and licenses, general provisions (whistleblower protection, citizens right to press legal charges against the government), and other funding to states. Standards have been set for most forms of water pollution, including particulates, chemicals, biologicals, biosolids from treatment plants, and thermal pollution. The EPA recognizes that not all sources of water pollution can be controlled or prevented. Municipalities, for example, may need to discharge rainwater that contains street debris, and few, if any, water treatment plants are 100% effective. Therefore, the EPA issues **National Pollutant Discharge Elimination System** (**NPDES**) permits that allow some discharges into public water sources.

12.3.6.4   Resource Conservation and Recovery Act (RCRA, 1976)

This is the principal federal law covering solid and disposable wastes. Solid waste includes all garbage or refuse; sludge from wastewater or water supply treatment plants; wastes from air pollution control facilities; and other discarded materials including solid, liquid, semisolid, or contained gaseous material resulting from industrial, commercial, mining, agricultural, and community activities (U.S. EPA, 2016c). The current law focuses on controlling the ongoing generation and management of solid waste streams in what's been called a cradle-to-grave oversight. This means that the EPA requires detailed information on solid wastes from their point of generation through their transportation, treatment, and storage to their ultimate disposal or destruction.

12.3.6.5   Toxic Substances Control Act (TSCA, 1976)

This law authorizes the EPA to require reporting, record-keeping and testing requirements, and restrictions relating to chemical substances and/or mixtures. Substances such as food, drugs, cosmetics, and pesticides are generally covered by other laws, but TSCA includes radon, polychlorinated biphenyls, asbestos, lead-based paint, and an ever-growing list of chemicals (U.S. EPA, 2016d). Any time a chemical company proposes to market a new chemical, it is mandated to contact the EPA for registration. Following strict protocols, the company provides toxicity testing that incorporates animal and plant models with a provision to look for chemicals that may be carcinogenic to humans. If the testing is accepted and the appropriate paperwork is filed, the EPA may issue its approval to pursue larger-scale manufacturing. The EPA also maintains the ever-growing TSCA Inventory, which currently lists more than 85,000 chemicals (U.S. EPA, 2016d).

Contrary to its title, the TSCA Inventory includes chemicals that range from low to very high toxicity, and the same regulations pertain to all. Approximately 62,000 chemicals were grandfathered in at the enacting of the TSCA, which means that if they existed before TSCA they did not need to go through toxicity testing. By law, these chemicals are not considered to impose 'unreasonable risk' even though we know that some are highly toxic based on independent testing. Many other chemicals are excluded from TSCA provisions, including those manufactured only in small quantities for research purposes and those covered by other federal legislation. Of the 85,000+ chemicals, only about 250 have been tested for their toxicity under the EPA as of 2015 (U.S. EPA, 2016d). There are 3000 high production volume chemicals produced or imported in quantities exceeding 1 million lb per year in that list.

12.3.6.6 Comprehensive Environmental Response, Compensation, and Liability Act (CERCLA, 1980)

The Comprehensive Environmental Response, Compensation, and Liability Act was passed to inventory and remediate the huge number of abandoned or poorly managed hazardous waste sites scattered throughout the county. If a hazard review of a site demonstrates that it is a significant problem, the site is registered on the EPA's **National Priorities List** (NPL) as a **Superfund Site**. The EPA also maintains a **Toxic Release Inventory** (TRI) of all prominent chemical spills in the country. Together, Superfund and TRI sites are extremely numerous, especially in the eastern half of the nation (Figure 12.6).

Priority for the remediation of Superfund sites is based on the score the site receives during a hazard review. At the end of 2016 there were 1337 Superfund site, 53 proposed sites and 392 that had been deleted and did not need further remediation (U.S. EPA, 2016e). When possible, the EPA identifies the potentially responsible party to take over the costs of cleanup. If no such party can be found, the EPA pays for cleanup through a special Superfund trust fund. After remediation, the site undergoes a 5-year observation period to assure that nothing pops up, literally or figuratively.

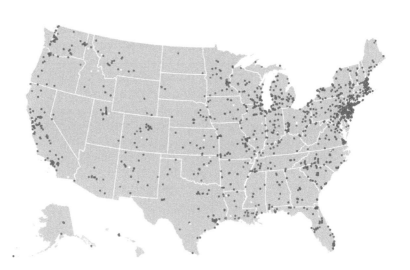

FIGURE 12.6   Map of the United States showing significant toxic releases and Superfund sites. (Courtesy of U.S. EPA, Learn about the Toxics Release Inventory, 2017, https://www.epa.gov/toxics-release-inventory-tri-program/learn-about-toxics-release-inventory.)

12.3.6.7 Federal Insecticide, Fungicide, and Rodenticide Act (FIFRA, 1996)

The TSCA does not cover pesticides, but this gap is filled by FIFRA. All pesticides distributed or sold in the United States must be registered by the EPA. Before the EPA allows a pesticide to be registered or approved for use, the applicant must show, among other things, that using the pesticide per specifications "will not generally cause **unreasonable adverse effects** on the environment." This requires that the manufacturer conduct studies on the effectiveness of the proposed pesticide, its chemical fate and transport, and its toxicity. If there is a risk that the chemical can cause human cancer, a more extensive battery of tests must be conducted.

After the testing has been conducted, the manufacturer may submit the information and request that the chemical be registered for use. The company must also prepare a label that describes how the pesticide is to be used, as well as other information about active ingredients, inert ingredients, and signs of toxicity. Regulation of home application of pesticides is rather lax, but commercial applicators must follow the product label exactly or they are subject to fines and loss of their applicator's license. The EPA maintains a large list of restricted-use pesticides that are not available to the general public because they pose too great a risk.

It costs millions of dollars to register a pesticide, so manufacturers are very particular about the chemicals that they intend to register. A frequent criticism is that the EPA allows the chemical company to conduct its own testing rather than having them submit the chemical to independent laboratories (Boone et al., 2014) and that may be akin to having the fox guard the chicken house.

The EPA is currently undergoing a detailed reregistering of pesticide products whose initial registration is close to expiring. Reregistration potentially involved more than 24,000 products, but over 10,000 of these were quickly taken off of the market either by the EPA or voluntarily by the manufacturer. In many of these cases the pesticide was no longer being used but had remained on the books.

12.4 REGULATION AT THE STATE AND MUNICIPAL LEVELS

Every state in the United States has an agency that oversees environmental quality. Some have their agencies as a subunit of their conservation departments, but most have separate agencies dealing with the state's laws concerning contaminants. Their objectives almost universally include monitoring, education, enforcement, and remediation.

States can establish their own pollution standards if they are not less rigorous than federal standards. California is one of the states that imposes stricter standards on many contaminants, especially pesticides, than the federal regulations. California also has the most complete database on pesticide sales and use among all other states.

Major cities also have municipal offices that issue permits, monitor air and water quality, and assess fines when necessary. Smaller cities may be members of regional consortiums that do the same. All local and regional authorities must have standards that at least comply with those established by the U.S. EPA.

12.5   CHAPTER SUMMARY

1. The regulation of contaminants and their sources requires a global effort. Regulatory agencies exist at the international, national, state, and municipal levels.
2. At the international level, the United Nations has its Environment Programme, the European Union contains the Directorate-General for the Environment. Whereas the United Nations cannot assign fines or penalties, the European Union can for the countries that are members.
3. Within the United States, many cabinet departments have some regulatory authority for pollution, even if it's only within their own lands. The Department of Defense and the Bureau of Land Management within the Department of the Interior are the two largest landholders in the federal government and consequently have their share of environmental problems. The Bureau, however, has a small budget relative to the millions of acres under its prevue and often has to rely on other agencies for help.
4. The U.S. Environmental Protection Agency is the principal federal watchdog over contaminants. It has been given authority over many federal acts that regulate contaminants including the Clean Water Act; Clean Air Act; Toxic Substances Control Act; Federal Insecticide, Fungicide, and Rodenticide Act; and the Comprehensive Environmental Response, Compensation, and Liability Act, to name a few.
5. In addition to the federal agencies, each state has its own agency that oversees and regulates the discharge of chemicals into the environment. By law, state regulations can be stricter but not more lenient than federal regulations. Many larger cities have bureaus or offices that oversee chemical usage and wastes in their jurisdictions.

12.6   SELF-TEST

1. True or False. The United Nations has authority to directly fine or otherwise penalize nations for not complying with international environmental agreements.
2. What does the acronym UNEP stand for?
   a. Unified National Environmental Program
   b. United Nations Environmental Policy
   c. United Nations Ecology Programme
   d. United Nations Environment Programme
3. Which of the following international agreements focuses on persistent organic pollutants?
   a. Stockholm Convention
   b. Kyoto Protocol
   c. Vienna Convention
   d. Montreal Protocol
4. True or False. Once a nation signs off on a United Nations protocol or convention it automatically becomes the law of that nation.

5. Which of the following international bodies can assess fines and other penalties on member countries if they violate accepted agreements?
   a. United Nations
   b. OECD
   c. European Union
   d. More than one above
   e. None of the above
6. Describe in your own words what FIFRA stands for and what does it cover?
7. What bureau or agency has the greatest responsibility for the welfare of migratory fish and wildlife in the United States?
   a. Environmental Protection Agency
   b. Bureau of Land Management
   c. U.S. Fish and Wildlife Service
   d. Department of Defense
8. Which of the following departments has no involvement with contaminant issues?
   a. Defense
   b. Commerce
   c. Agriculture
   d. Interior
   e. All have at least some involvement with contaminant issues.
9. What is a Superfund site? What agency has responsibility for these sites?
10. Both the Toxic Substances Control Act (TSCA, 1976) and the Federal Insecticide, Fungicide, and Rodenticide Act (1996) deal with chemical contaminants. How do the two Acts distinguish authorities? That is, how do they differ with regard to the chemicals they regulate?

REFERENCES

Boone, M.D., Bishop, C.A., Boswell, L.A., Brodman, R.D., Burger, J. et al. 2014. Pesticide regulation amid the influence of industry. *BioScience* 64: 917–922.

Bureau of Safety and Environmental Enforcement (BSEE). 2016. About us. https://www.bsee.gov/who-we-are/about-us.

European Commission. 2016. Environment. http://ec.europa.eu/dgs/environment/index_en.htm. Accessed November 28, 2016.

Organisation for Economic Co-operation and Development (OECD). 2016. OECD home. http://www.oecd.org/about/. Accessed November 28, 2016.

United Nations Environment Programme (UNEP). 2016. Stockholm convention. http://chm.pops.int/default.aspx. Accessed November 28, 2016.

U.S. Environmental Protection Agency (EPA). 2016a. Persistent organic pollutants: A global issue, a global response. https://www.epa.gov/international-cooperation/persistent-organic-pollutants-global-issue-global-response. Accessed November 28, 2016.

U.S. Environmental Protection Agency (EPA). 2016b. Section 404 of the clean water act: How wetlands are defined and identified. https://www.epa.gov/cwa-404/section-404-clean-water-act-how-wetlands-are-defined-and-identified. Accessed November 28, 2016.

U.S. Environmental Protection Agency (EPA). 2016c. Resource Conservation and Recovery Act (RCRA) laws and regulations. https://www.epa.gov/rcra. Accessed November 28, 2016.

U.S. Environmental Protection Agency (EPA). 2016d. TSCA toxic substances inventory. https://www.epa.gov/tsca-inventory.

U.S. Environmental Protection Agency (EPA). 2016e. Summary of the Comprehensive Environmental Response, Compensation, and Liability Act (Superfund). https://www.epa.gov/laws-regulations/summary-comprehensive-environmental-response-compensation-and-liability-act. Accessed November 28, 2016.

U.S. Environmental Protection Agency (EPA). 2017. Learn about the Toxics Release Inventory. https://www.epa.gov/toxics-release-inventory-tri-program/learn-about-toxics-release-inventory.

13 Future Perspectives and Concluding Remarks

13.1 INTRODUCTION

Since this is the last chapter in this book (do I hear shouts of joy from some students?), it is by tradition that I am supposed to reminisce a bit and then unwrap my crystal ball and predict the future of ecotoxicology—that is, where the science is headed. Alright, I'll try to do my best in this regard but don't blame me if some new developments occur in the science that I did not see forthcoming.

13.2 DOES LOOKING BACKWARD TELL US ANYTHING ABOUT WHERE THE SCIENCE IS HEADED?

The answer, I think, is yes—a retrospective on the science of ecotoxicology does reveal some things about the near future at least. The science began with a sincere interest in how chemical contaminants can affect fish and wildlife. As we have seen, emphasis was on acute toxicity but that evolved into a deeper appreciation for the many sublethal effects that contaminants cause and eventually how those physiological effects might harm higher levels of ecological organization including populations and communities. These latter concerns are still pretty much in their infancy, however. Around the same time, environmental chemists enthusiastically began collecting environmental samples and analyzing them for their contaminants. Gradually, they built up sufficient data to make predictions on how chemicals behave in natural situations that led to the more refined understandings we have today. A lot has happened in a relatively short period of time.

One way of studying the history of a science is to look at its publication record. Search engines such as Web of Science™ greatly help in this effort. If we only look at pesticides, we find that the first paper was published in the late 1800s (*Science*, 1889). At that time one author thought that arsenic-based compounds such as Paris green and London purple were terrific. Today we recognize that these chemicals can just as easily kill humans as they do insects. He listed less than 10 compounds that growers used to control pests, that's all. He also mentioned that annual crop losses to insects was between $200 and $250 million across the United States. That would be about $5 to $6.2 billion today.

In 1943, an article in *Science* cited 5000 different compounds used as insecticides or fungicides (Frear, 1943). Many of these were homemade remedies, however. DDT was not mentioned then, but 2 years later, over 200 scientific papers were published on this insecticide, only a few of which mentioned concerns; most papers praised the chemical for its ability to control insects. Other insecticides included a few naturally

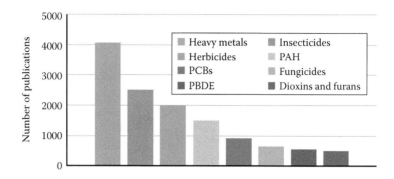

FIGURE 13.1   Number of scientific articles published by major contaminant group in 2016. (Courtesy of Web of Science.)

derived compounds, pyrethrins, oils, rotenone, and a couple of other organochlorines. Organophosphorus and carbamate pesticides did not make the scene until 1965.

About that time, research interest in pesticides exploded (refer to Figure 13.1). The number of research papers on pesticides has continued to grow exponentially through today. Other organic contaminants that were known at that time, such as polycyclic aromatic hydrocarbons, showed a similar trend, although not as domineering as pesticides. Polychlorinated biphenyls were of little interest until the mid- to late 1970s, while plastics and personal care products did not make it on the scientific radar screen until several years later. Heavy metals have been of major concern, and due to the number of metals that can be included in this group, they have been well represented in the literature.

The late 1960s and 1970s were a period of public awakening to environmental problems. Certainly, Rachel Carson's book *Silent Spring* (1962) was instrumental in this interest, but such movements tend to be self-feeding. Authors see an increase in public interest and write books that feed this interest, which causes greater awareness, etc., until eventually the interest wanes. At the same time, several key pieces of federal legislature were passed and significant federal budgeting for research on the environment suddenly appeared. This stimulated research on water, air, and terrestrial environmental science. Once the foundations for ecotoxicology were laid, the infrastructure rose quickly with the institution of the Environmental Protection Agency (EPA) and most state environmental agencies. An ecological awareness eventually became global with the United States, India (surprisingly??), Canada, and the United Kingdom as its principal activists.

Over the years, there has been a shift in interest as attested by funding of research. Initially, ample funding at the federal level was on the effects of contaminants on the environment, including wildlife and fish. Whether the research was done in federal research centers or in funded university and corporate laboratories, substantial information was obtained on the toxicity of many chemicals on a handful of species. Compared to the number of known chemicals and species, however, we have only been able to scratch the surface. Understandably, a very large proportion of research was conducted on traditional lab animals, with human health as a primary concern.

Funding for analyzing environmental samples like water, sediment, soils, and air also became available from several sources, especially as regulatory agencies demanded it as part of risk assessment. It is hard to say when it occurred because the change was gradual, but over time emphasis increasingly went toward funding chemical analyses until that overtook funding of effects studies. It seemed that funding agencies, particularly regulatory agencies such as the EPA, came to tacitly believe that the really important need was to find, characterize, and remediate contaminated sites. Certainly, some effects data were available from manufacturers when they wanted to register a new product, but those data were kept secret and came from rarified lab studies on species that might have no relevance to the contaminated site. I must admit, I do not have actual data on how much funding actually goes into quantifying contaminants and how much goes into effects, so I may be biased, but it may also be time to take a look at such overriding issues. There are many facts to support this contention.

13.3   WHAT IS THE CURRENT STATUS OF ECOTOXICOLOGY?

Despite the perception of reduced funding, ecotoxicology remains a very healthy science. In 2016 alone, thousands of scientific papers were published on contaminants in the environment (Figure 13.1). Not all of these numbers represent unique papers, however, because many papers discuss multiple chemicals and would reappear with more than one keyword. Nevertheless, the numbers are still impressive.

The science of ecotoxicology has produced significant benefits for other sciences. In ecology we have shown that chemicals in the environment can be viewed as any of the other stressors affecting ecological receptors such as individuals, populations, and habitats. Consequently, scientists have developed an arsenal of tools to explore these stressors. In toxicology, scientists have demonstrated over the years that it is the sublethal effects that are most relevant in natural environments. Thus, in addition to refining the concepts of LC50 and LD50, ecotoxicologists have provided data on genotoxicity, immunosuppression, endocrine disruption, and other maladies organisms suffer from contaminants. In physiology, ecotoxicologists have not only shown that such effects can occur but the mechanisms through which they are affected. Even in the area of public law and regulation, findings from contaminant studies have improved risk assessment and led to important laws that protect our environment from further damage. These are just some of the ways that ecotoxicology has contributed to other areas of science.

13.4   WHERE SHOULD THE SCIENCE HEAD?

If you asked 100 ecotoxicologists this question, you would probably receive 200 different responses. So, this is one person's thoughts on this topic. I'll try to highlight a few areas that I think are important. This might be a good topic to discuss during the final week of classes, if you have time.

If I ruled the Kingdom of Ecotoxicology, I would increase funding into a few needed areas while still maintaining other areas. These are not new areas, for some work has been occurring for over a decade or two, but the amount of work compared to the need for understanding is alarming, at least in my humble opinion.

13.4.1 CONDUCT STUDIES ON THE HUGE INVENTORY UNDER THE TOXIC SUBSTANCES CONTROL ACT (TSCA)

There are tens of thousands of chemicals on the TSCA registry (see Chapter 11). Only a very few of these have had any toxicity testing at all. There have been efforts to model prospective toxicity by examining the structure of some of these molecules to derive some sort of toxicity reference value (TRV), but it is my understanding that these models need more validation over a broad range of chemicals and species. Perhaps the new Frank R. Lautenberg Chemical Safety for the 21st Century Act (EPA, 2016a) may facilitate this by prioritizing existing and new chemicals for risk evaluations, but the recommended rate of new assessments (20 in 3.5 years) is woefully inadequate.

13.4.2 EFFECTS OF MULTIPLE CONTAMINANTS ON ORGANISMS

We have quite a bit of information on the effects of single contaminants on several species of plants and animals. However, we also know that mixtures of chemicals may have very different results than a simple adding up of individual toxicities. Interactions among contaminants may be additive ($1 + 1 = 2$), synergistic (the total is greater than the parts or $1 + 1 > 2$), inhibitory ($1 + 1 < 2$), or something totally different ($1 + 1 = A$). When organisms are subjected to 2, 5, 10, or more contaminants simultaneously, all sorts of things can happen, or not. The excitement lies in seeing what can happen. Studying multiple chemicals on organisms requires substantial funding and space. Let us provide a simple example. Suppose a scientist wishes to determine the effects of two chemicals on a species with proper controls. A full study would require a battery of test units for each chemical plus another battery of units for the mixtures. Each battery would have a set of concentrations and replicates at each concentration. The number of experimental units increases geometrically with each added chemical.

13.4.3 INCREASED CONSIDERATION BY REGULATORY AGENCIES FOR RELEVANT SPECIES

The EPA and comparable state agencies have long used a somewhat short list of preferred species for their registration and regulatory processes. While these species may be relatively easy to maintain and even breed under laboratory conditions and while there is a large volume of proprietary and published studies on these species, they are not always relevant to the particular risk assessment or problem under study. For a long while, for example, the EPA would frown on studies that used amphibian larvae, even though amphibians, not fish, were potentially at risk. Similarly, a lot of focus has been on northern bobwhite (*Colinus virginianus*), Coturnix quail (*Coturnix coturnix*), ring-necked pheasant (*Phasianus colchicus*), mallard (*Anas platyrhynchos*) and laboratory mammals, but do they suffice for all birds and mammals?

13.4.4 ALTERNATIVE METHODS

At the same time that more relevant species are being considered, concern for the welfare of animals requires that alternative methods to traditional toxicity testing with live animals be entertained when possible. The EPA, for example, is well aware of this and has published guidelines for consideration (EPA, 2016b). Some of these methods include modeling, use of tissue banks, and alternative statistical analyses. All of these methods would need extensive validation to make certain that they are reliable and representative of real organisms. Increased interest in developing these methods should be encouraged.

13.4.5 Develop Realistic Scenarios

As I have mentioned several times during this book, laboratory studies are essential for some questions but lack the realism that organisms face in their native habitats. Natural areas vary through time, sometimes on a daily basis, which makes finding a good, clear result based on toxicity testing difficult. However, these are the environments in which organisms occur, where the real world occurs. Very well-controlled studies under close confinement cannot mimic natural exposures. Funding should go toward studies that seek how to bridge the need for precise toxicological data with real environmental study designs. We can build on existing frameworks such as mesocosms and larger-scale holding facilities.

13.4.6 Increase Study of Higher Level Effects

Chapter 9 and Chapter 10 should have made it clear that while we have some concepts of how contaminants affect higher level ecological receptors such as populations, communities, and ecosystems, there is a whole lot more we could learn. In some cases we can do this experimentally; in other instances we may have to rely on weight-of-evidence types of studies. The public uncertainty about global climate change is a reflection of how difficult some of these studies may be. The two types of uncertainty have tendencies to increase as the complexity increases.

13.4.7 Gain More Information on Nanoparticles and Their Effects

Over the years scientists have come to understand the chemical fate and transport and the effects exerted by most of the contaminants discussed in this book. Certainly, there are unanswered questions and research will continue to address these questions. However, the one category of contaminants that should be of growing concern is nanoparticles. As explained in Chapter 8, it appears that we have just tapped the leading edge of nanotechnology. Over the next several years, application of nanos is bound to grow tremendously. Because nanos behave in ways that differ from their larger particles, ways that include both physiological and environmental effects, ecotoxicologists will have a difficult time keeping pace with their development. We do not want to look back several years from now and say "If only …."

13.4.8 Encourage Advancements in Environmental Chemistry and Risk Assessment

Of course, environmental chemistry will continue to be a very active and productive science as manufacturers produce new generations of chemicals. Methods to detect and quantify these chemicals in many different ecological receptors at ever increasing sensitivity will be demanded in risk assessments. Risk assessment and management, too, need continuous updating and improvements to reduce waste and improve decision making.

REFERENCES

Carson, R. 1962. *Silent Spring*. Boston, MA: Houghton Mifflin Harcourt, 400pp.
Environmental Protection Agency (EPA). 2016a. The Frank R. Lautenberg Chemical Safety for the 21st Century Act. https://www.epa.gov/assessing-and-managing-chemicals-under-tsca/frank-r-lautenberg-chemical-safety-21st-century-act. Accessed December 29, 2016.
Environmental Protection Agency (EPA). 2016b. Process for establishing & implementing alternative approaches to traditional in vivo acute toxicity studies. https://www.epa.gov/pesticide-science-and-assessing-pesticide-risks/process-establishing-implementing-alternative. Accessed December 30, 2016.
Frear, D.E.H. 1943. A catalogue of insecticides and fungicides. *Science* 98: 585.
Science. 1889. Insecticides and their application. *Science* 13: 393–396.

Glossary

Following is a list of terms and their definitions used in this book. It does not include all of the words highlighted in bold in the text because many of those are common words or phrases that were emphasized. This list contains many terms that are important to the understanding of ecotoxicology but may be less familiar to students.

Acetylcholine: A neurotransmitter commonly found in the neuromuscular junctions and in the brain.

Acetylcholinesterase: An enzyme that deactivates acetylcholine. Organophosphorus and carbamate pesticides deactivate the enzyme, which causes repetitive firing of neurons.

Acute toxicity: Toxicity or poisoning that occurs over a short period of time, usually hours to 4 days in toxicity tests.

Adduct: An attachment to DNA or RNA that disrupts protein production. Some contaminants act as adducts.

Aleatory uncertainty: The type of uncertainty in science that is caused by random variation in the system being studied.

Anthropogenic: Caused or produced by human activities.

Apoptosis: Cellular death. It can be natural or induced by chemicals.

Aroclor: During the marketing of PCBs, the manufacturer combined several PCBs into mixtures called Aroclors based on the degree of chlorination.

Bay region: A structural region on some PAHs that facilitates the formation of adducts that can disrupt DNA and protein synthesis.

Bioassimilation: The uptake of contaminants by organisms and incorporation into their cells.

Bioconcentration: Bioassimilation that leads to higher concentration of contaminants in organism tissues than in the environmental medium that served as a source of the contaminants.

Biomagnification: When contaminants, mostly lipophilic ones, are passed up the food chain to the highest trophic levels; concentrations increase at highest trophic levels.

Carbamate pesticide: A type of 'current use' pesticide formed from carbamic acid. Carbamates reversibly deactivate acetylcholinesterase in the motor neuron synapses and can lead to death usually through asphyxiation.

Chronic toxicity: Toxicity that occurs over a prolonged period of time. In chronic toxicity tests, organisms may be exposed to contaminants for periods lasting a week to several months.

Circumneutral: Deals with pH. If soils or water have a pH between 6.5 and 8.0 (some variation in these limits), they are said to be circumneutral or around neutral pH of 7.0 and safe to most organisms.

Congener: An individual molecular type of polychlorinated biphenyls, polybrominated biphenyls, or polybrominated diphenyl ethers. There are 209 possible congeners in each molecular family.

Contaminant: Any substance that reduces the purity of another substance such as water, air, or soil.

Coplanar: Polychlorinated biphenyls in which the two phenyl groups lie in the same physical plane. Coplanar PCBs are usually more toxic than nonplanar PCBs.

Depuration: The loss of a chemical substance from an organism through excretion, urination, or evaporation. Sometimes the decomposition of chemicals into simpler molecules is considered part of depuration if degradation facilitates removal from the body.

Dioxin-like: Dioxins, PCBs, and furans that have elevated toxicity compared to other chemicals in their family due to structural similarities with 2,3,7,8-tetrachlorodibenzo-para-dioxin (TCDD).

Ecotoxicology: The science that studies contaminants in the environment including the fate and transport of chemicals and their effects on ecological receptors including organisms and habitats.

Endocrine disruption: Interference with normal endocrine functioning through exposure to contaminants. These endocrine disrupting chemicals can enhance or block the action of hormones. The greatest amount of study has been on sexual and thyroid functioning.

Epistemic uncertainty: The type of uncertainty in science that is caused by inadequate study design. It may be corrected through increased sampling effort and other methods unlike aleatory uncertainty, which cannot be easily corrected.

Genotoxicity: Interference with the genetics of an organisms through contaminant exposure. Possible effects include damage to chromosomes through breakage or alterations in the sequencing patterns of DNA such as insertions, deletions, or substitutions.

Half-life: The time required for one half of chemical substance to degrade. It may refer to radioactive materials and their loss of neutrons or the decomposition of complex, long-lasting contaminants. Half-lives are constants regardless of the amount of material still present.

Halogenated: Organic molecules that have at least one halogen atom attached to them. Common halogens include chlorine, bromine, and fluorine.

Hazard: In risk assessment, a hazard is something that can cause or contribute to risk. A contaminant in water, for example, may be a hazard but so can a huge pile of boulders.

Hazard quotient: This is the ratio between the concentration or strength of a hazard and the level at which it can pose a risk. $HQ = (Site\ Exposure\ Level)/TRV$ where HQ is the hazard quotient and TRV is the toxicity reference value. If $HQ > 1$, then risk is presumed; if it is <1, then no or very limited risk is likely.

Heavy metals: By definition, heavy metals are elements falling within one of the metal groups in the Periodic Chart of Elements that have a density greater than iron. In practice, within ecotoxicology heavy metals are dense metals that are sufficiently abundant and toxic to cause concern. They include cadmium, chromium, copper, lead, mercury, and zinc, but occasionally other metals are also included in this group.

Hydrophilic: Organic molecules that dissolve or mix readily with water. This is the antonym of hydrophobic, which are molecules that do not readily mix with water.

Hyperaccumulator: An organism, usually a plant, that can uptake large concentrations of certain contaminants.

Immunotoxicology: The study of how toxins such as contaminants can affect the immune system of organisms.

Keystone species: Species whose ecological importance is greater than what would be expected based on abundance or mass alone. Keystone species have important roles in maintaining ecological homeostasis.

LC50: The median lethal dose or the concentration of a toxin that will kill 50% of a sample population when administered directly to the animal through injection, gavage, or similar method.

LD50: The median lethal dose or concentration of a toxin when placed in food or other method where the concentration is not well known.

Legacy compounds: Persistent organic compounds such as organochlorine pesticides that were used in the past but not currently and are still in the environment due to their persistence.

Lethal dose: The amount of a toxin such as a contaminant required to kill an individual organism.

Lipophilic: Organic chemicals that mix well or dissolve in lipids or organic solvents.

Median lethal dose: Similar to the LC50 or LD50. This is the amount of a toxin necessary to kill 50% of a sample population.

Organometals: Metals that are combined with organic molecules such as methylmercury or selenomethionine. The organic moiety increases assimilation rates by organisms, and often the combination is more toxic than either element alone.

Organophosphate pesticide: A group of 'currently used' pesticides noted for their ability to permanently deactivate acetylcholinesterase and thereby cause continual firing of neurons.

Persistent organic pollutants: Chemicals identified by the United Nations Stockholm Convention as being long lasting in the environment. They include legacy compounds and those that were made more recently.

Pyrethroid: A group of pesticides that were originally developed from the chrysanthemum flower but are now almost entirely synthetic. Their mode of action is to prevent closure of voltage-gated sodium channels in the axons of neurons, resulting in continual firing of the neuron.

Risk assessment: The process of determining which, if any, ecological receptors are at risk from disturbances in the environment. In the context of this book, risk assessment evaluates the actual and potential damage caused by contaminants.

Risk management: Once risk assessment has identified threats or possible harm due to contaminants, the process of risk management attempts to identify the most efficacious way of treating that risk in a manner that balances ecological damage, financial costs, and the legal or regulatory agencies.

Sublethal dose: The concentration of a toxin such as a contaminant that is insufficient to cause direct mortality but may exert other effects such as reduced growth, endocrine disruption, genotoxicity, or a host of other harms.

Superfund: A contaminated site that falls under the prevue of the Comprehensive Environmental Response, Compensation, and Liability Act.

Teratogenic: A toxic compound that causes malformations in organism before they are born or hatched.

Toxicity reference value: An estimate of the total toxicity of an environmental source such as water, air, or soil determined from the kinds and concentrations of contaminants present and known toxicity of these compounds to animals or plants.

Vulnerability analysis: The process of determining how sensitive an ecological receptor is to environmental stressors. With regard to organisms, vulnerability analysis may occur as an acute or sublethal toxicity test.

Index

A

Acceptable risk limits, 184
Accipiter cooperii, 117–118
Acetylcholine (ACh), 32–33
Acetylcholinesterase (AChE), 32–33, 70
Acid neutralizing capacity (ANC), 133
Additive mortality, 143–144
Adducts, 36
Aleatory uncertainty, 181
Alligator mississippensis, 54
α-cypermethrin, 155
Ambystoma tigrinum, 108
Ameiurus nebulosus, 29, 107
Anas platyrhynchos, 39, 93
Anthropogenic sources, 3
Apoptosis, 118, 161
Arbuscular mycorrhizae (AM), 165
Aroclors, 87
Arsenic (As), 16, 50–51, 58
Asio otus, 116
Atomic mass, 11–12
Atomic number, 11
Atrazine, 76–77

B

Bacillus thuringiensis (*Bt*), 78, 194
Benzo[a]pyrene (BaP), 106–107
Biological concentration factors (BCFs), 105–106
Biphenyl, 84–85
Bisphenol A (BPA), 35, 128, 149
Brevoortia patronus, 118
Bureau of Land Management (BLM), 195
Bureau of Ocean Energy Management (BEOM), 196
Bureau of Safety and Environmental Enforcement (BSEE), 196

C

Calcium carbonate (CaCO₃), 50
Carbamates, 69–70
Caretta caretta, 28–29
Carson, Rachel, 5
Cathartes aura, 116
Cavia sp., 93
Centers for Disease Control (CDC), 52–53
CERCLA, *see* Comprehensive Environmental Response, Compensation, and Liability Act

Chelydra serpentina, 54, 147
Chemical balance equation, 180
Chronic toxicity, 3–4
Chrysanthemum coccineum, 71–72
Clean Air Act, 198
Clean Water Act, 198–199
Colchicus phasianellus, 208
Colinus virginianus, 39, 93, 208
Community
    ecological succession, 166–168
    food webs/chains, 163–165
    level of evenness, 162
    meiofauna, 163
    sensitivity, 168
    species diversity, 162–163
    species richness, 161–163
    symbiotic relationships, 165–166
Compensatory mortality, 143–144
Competitive binding, 50
Complement, 34
Comprehensive Environmental Response, Compensation, and Liability Act (CERCLA), 200
Congeners, 85
Contaminants
    acid deposition
        and elevated aluminum, 135–136
        embryonic amphibians, 135
        Midwestern soils, 134
        nitrogen oxide, 132
        pH of water, 134–135
        smokestacks, 133
        sulfur dioxide, 132–133
    community
        ecological succession, 166–168
        food webs/chains, 163–165
        level of evenness, 162
        meiofauna, 163
        sensitivity, 168
        species diversity, 162–163
        species richness, 161–163
        symbiotic relationships, 165–166
    ecosystem dynamics
        climate change, 170–171
        D. pulex populations, 170
        marine and estuarine habitats, 170
        MeHg, 171
        nutrient and mineral balances, 169–170
        photosynthesis, 168
        primary productivity, 169

respiration, 168
secondary productivity, 169
multiple chemicals, 208
nanoparticles, 209
carbon atoms, 129–130
glazes, 130
health hazards, 132
industrial applications, 130
nano liposomes, 130–131
nanotubules, 130
toxicity, 131–132
national regulation
BLM, 195
BSEE, 196
crosscut governments, 192–193
Department of Agriculture, 194–195
Department of Commerce, 196–197
Department of Defense, 194
fish and wildlife resources, 195
oil spills, 192
state and municipal levels, 201
State Department, 193–194
U.S. EPA, 197–201
USGS, 196
pharmaceuticals, 128–129
plastics
biodegradable qualities, 124–125
BPA, 128
composition, 124
discarded/lost fishing nets, 125–126
environment, 125
EPA, 123
phthalates, 125, 127–128
plastic bags and jellyfish, 125, 127
polyethylene and polystyrene, 123–124
soda ring, 125, 127
subtropical convergence zone, 125–126
toxic chemicals, 125
populations
abundance, 146–148
additive and compensatory mortality,
143–144
age structure, 150–153
density dependent and independent
factors, 145–146
frogs, California, 144
life table, 153–156
sex ratios, 148–150
public law and regulation findings, 207
uncertainty, 209
Coturnix coturnix, 208
Coturnix japonica, 60
Covalent bonds, 15
Crassius auratus, 117
Cricetus sp., 93
Criteria pollutants, 198
Cucurbita pepo, 90

Cyclodienes, 110
Cygnus olor, 55
Cyprinus carpio, 132
Cytochrome P450, 6–37

D

Danio rerio, 93, 149
Daphnia sp., 37–38
D. *magna*, 90, 105
D. *pulex*, 105, 168
D. *schoedleri*, 155
Diastereomers, 113
Dichlorodiphenyltrichloroethane (DDT), 35,
110–113
Dichlorodiphenylchloroethylene (DDE), 111–113
Diclofenac, 129
Dioxins, 35
accidental fires/breakdowns, 91
aryl hydrocarbon receptor, 92
characteristics, 91–92
properties, 84, 91
sugarcane, 90–91
TCDD, 92–93
TEF, 92–93
TEQ, 92–93
Directorate-General for the Environment, 192
Dissolved organic carbon (DOM), 50
Double bond, 15
Dry deposition, 132–133
Dry weight, 23

E

Earth Day, 5–6
Ecological risk assessment
aquatic environments, 176–177
contaminants, 177
cost-benefit analysis, 178
decision making, 178
definition, 175
process, 177
Ecological Services Branch, 195
Ecosystem dynamics
climate change, 170–171
D. *pulex* populations, 170
marine and estuarine habitats, 170
MeHg, 171
nutrient and mineral balances, 169–170
photosynthesis, 168
primary productivity, 169
respiration, 168
secondary productivity, 169
ED, *see* Endocrine disruption
Elements
atomic mass, 11–12
atomic number, 11

chemical reactions, 14–15
energy shells, 12
flerovium, 11
metals
    in environment, 18
    heavy metals, 16
    metalloids, 16, 22
    properties, 15–16
organic molecules
    carbon–halogen bonds, 19
    contaminants, 20
    electromagnetic spectrum, 19–20
    hydrocarbons, 16–17
    isomers, 16–17
    octanol/water partitioning, 20–21
    phenol, 17
    rate of degradation, 18–19
    saturated molecules, 17
    soil organic carbon/water partitioning
        coefficient, 21
    weathering processes, 20
organochlorine pesticides, 21–22
PAH, 21
PCBs, 21
periodic chart, 12–14
plastics, radionuclides and munition
    compounds, 22
radioactive elements, 18
units of measurement, 22–23
valence shell, 12
Elodea sp., 37–38
Emigration, 147, 148, 157
Enantiomers, 113
Endocrine disruption (ED), 34–35, 106,
    118–119, 177
Energy shells, 12
Enhydra lutris, 109
Environmental chemistry, 209–210
Environmental Protection Agency (EPA), 53–54,
    101–102, 176
Epistemic uncertainty, 181
Esox lucius, 105–106
European Union (EU), 192–193
Extirpation, 161
Exxon Valdez spill crude oil, 109

F

Father of Toxicology, 5
Federal Food, Drug, and Cosmetic Act
    (FFDCA), 198
Federal Insecticide, Fungicide, and
    Rodenticide Act (FIFRA), 67,
    176, 201
Fresh weight, 23
Funding, 207, 209
Fungicides, 205–206

Furans, 35
    accidental fires/breakdowns, 91
    aryl hydrocarbon receptor, 92
    characteristics, 91–92
    properties, 84, 91
    sugarcane, 90–91
    TCDD, 92–93
    TEF, 92–93
    TEQ, 92–93

G

Gammaaminobutyric acid (GABA), 33
Gavia immer, 55
Genotoxicity, 35–36
Glufosinate, 75
Glyphosate, 74–76
Gymnogyps californianus, 56
Gyps sp
    G. bengalensis, 129
    G. indicus, 129
    G. tenuirostris, 129

H

Half-life, 3–4, 11, 18, 75, 86, 104, 113
Haliaeetus leucocephalus, 88
Halogens, 12
    compounds, 83–85
    dioxins and furans
        accidental fires/breakdowns, 91
        aryl hydrocarbon receptor, 92
        characteristics, 91–92
        properties, 84, 91
        sugarcane, 90–91
        TCDD, 92–93
        TEF, 92–93
        TEQ, 92–93
    fluorinated organic compounds, 94–96
    PBBs, 93–94
    PBDEs, 93–94
    PCBs
        Aroclors, 87
        biological effects, 90
        biphenyl, 84–85
        breakdown of, 87–88
        characteristics, 86
        concentrations, 88–90
        congeners, 85
        coplanar/nonplanar, 86
        dioxin-like, 85
        ortho-substitutions, 86
        persistence, 86–87
Hazard quotient (HQ), 180–181
Heavy metals, 16
Hexachlorocyclohexane (HCH), 111, 149
High-density polyethylene (HDPE), 124

Hyalella sp., 37–38
  H. azteca, 171
Hydrocarbons, 16–17
Hyperaccumulators, 51

I

Immigration, 147, 148, 157
Immunotoxicology, 34
Inert/noble gases, 14
Insecticides, 9, 69, 71, 73–74, 78–79,
        205–206
International Convention for the
        Prevention of Pollution, 197
International Court of Justice, 190
Invisible camouflage, 130
Ionic bonds, 14
Isomers, 16–17

J

Juncus roemerianus, 167

K

Kesterson National Wildlife Refuge, 30
Keystone species, 164
Kyoto Protocol, 190–191

L

LC50, 37–39
LD50, 38–39
LDPE, *see* Low-density polyethylene
Lead (Pb)
    acid water, 53
    aquatic biota, 55
    CDC, 52–53
    concentration of, 53
    environmental problem, 55
    mammalian toxicity, 56
    oxides and sulfides, 53
    plants and invertebrates, 54–55
    secondary poisoning, 55–56
    soil particles, 53–54
    uses, 53–54
    vertebrates, 54–55
Lethal dose, 4
Life table
    age distribution, 153
    analyses, 153
    Capitella sp., 156
    cohort life table, 153–155
    dynamic life table, 153
    PBDE, 155

reference population, 155
  Streblospio benedicti, 156
Live weight, 23
Low-density polyethylene (LDPE), 124
Lumbriculus variegatus, 37–38
Lutra canadensis, 148
Lymphokines, 34

M

Median lethal dose, 37
Mercury (Hg)
    amphibians and reptiles, 59
    birds and mammals, 59–60
    concentrations, 58–59
    Daphnia sp., 59
    dental amalgams, 56
    industry, medicine, and agriculture, 58
    land/water, 57
    MeHg, 56
    mercury cycle, 57
    methylated/ethylated, 56
    Minamata Disease, 60
    ocean sediments, 58
    organomercury, 58
    retention time cycle, 58
Mesocosms, 39–40
Metalloids, 16, 22–23, 47, 50
Metallothionein (MT), 51
Metals
    biological effects, 50–52
    in environment, 47–49
    factors, 49–50
    lead
        acid water, 53
        aquatic biota, 55
        CDC, 52–53
        concentration of, 53
        environmental problem, 55
        mammalian toxicity, 56
        oxides and sulfides, 53
        plants and invertebrates, 54–55
        secondary poisoning, 55–56
        soil particles, 53–54
        uses, 53–54
        vertebrates, 54–55
    mercury
        amphibians and reptiles, 59
        birds and mammals, 59–60
        concentrations, 58–59
        Daphnia sp., 59
        dental amalgams, 56
        industry, medicine, and agriculture, 58
        land/water, 57
        MeHg, 56
        mercury cycle, 57
        methylated/ethylated, 56

Minamata Disease, 60
ocean sediments, 58
organomercury, 58
priority pollutants, 52
Methylmercury (MeHg), 56, 171
Micropterus salmoides, 117, 135
Minamata disease, 60
Mono-ortho substituted PCBs, 86
Montreal Protocol, 191
Mortality, 147
MT, *see* Metallothionein
Mustela vison, 90
Mytilus edulis, 108

N

Nanoparticles, 209
carbon atoms, 129–130
glazes, 130
health hazards, 132
industrial applications, 130
nano liposomes, 130–131
nanotubules, 130
toxicity, 131–132
Natality, 147
National Marine Fisheries Service (NMFS), 195
National Oceanic and Atmospheric Administration
(NOAA), 196–197
National Pollutant Discharge Elimination System
(NPDES), 199
National Priorities List (NPL), 200
National Research Council (NRC), 176
Natural Resource Damage Assessment (NRDA), 195
Natural Resources Conservation Service (NRCS),
194–195
Non-ortho substituted PCBs, 86
Nonsteroidal anti-inflammatory drugs (NSAIDs),
129
Notophthalmus viridescens, 108

O

Odocoileus hemionus, 60
Odontesthes hatcheri, 117
Oncorhynchus
O. *kisutch*, 93
O. *mykiss*, 37–38, 105
Orcinus orca, 109
Organisation for Economic Co-operation and
Development (OECD), 191
Organochlorine pesticides (OCPs), 109
in animals, 115–117
birds/mammals, 117–118
cyclodienes, 110
DDT, 110
endocrine disruption effect, 118–119
in environment, 114–115

persistence, 112–114
physical properties, 111–112
potential effects, 119
sources and uses, 111
structure of, 111–113
vinyl chloride, 110
Organometals, 49
Organophosphates (OPs), 70–72
Oxidative stress, 36

P

Paresthesia, 74
Parus major, 135
PBBs, *see* Polybrominated biphenyls
PBDEs, *see* Polybrominated diphenyl
ethers
PCBs, *see* Polychlorinated biphenyls
Perfluoroalkyl substances (PFASs), 84–85,
95–96
Perfluorocarbons (PFCs), 84, 95
Perfluorooctane sulfonate (PFOS), 95–96
Perfluorooctanoic acid (PFOA), 84, 96
Periodic Chart of Elements, 12–14
Persistent organic pollutants (POPs), 149–150,
190–191
Pesticides
biologics, 78
carbamates, 69–70
chemical, 66
economic savings, 67–68
indiscriminate spraying, 67
inorganics, 77–78
nontarget organisms, 69
organic compounds, 78
organophosphates, 70–72
phosphonoglycine, 74–76
publications, 6
pyrethroids
birds and mammals, 74
half-lives, 73–74
insect toxicity, 73
paresthesia, 74
pyrethrum, 71–72
respiratory failure, 74
synthetic pyrethroids, 71, 73
target organism, 68
triazines, 74, 76–77
Pharmaceuticals, 35
Pharmaceuticals and personal care
products (PPCPs), 128–129
Phenol, 17
Phoca vitulina, 152
Phoebastria immutabilis, 28
Phosphonoglycine, 74–76
Phthalates, 35, 125, 127–128
Pimephales promelas, 37–38

Plants and animals
  cytochrome P450, 6–37
  endocrine disruption, 34–35
  endosulfan, 27
  enzymes
    chelation, 31
    contaminants, effect of, 33
    neurotransmitter interference, 32–33
    receptor binding, 31–32
  genotoxicity, 35–36
  immune system, 33–34
  laboratory exposures, 39–40
  LC50 test, 37–39
  LD50 test, 38–39
  lead/copper, 27
  malformations, 29–31
  oxidative stress, 36
  physical blockage, 28–29
  protocol, 37
  signs, 27
Plastics
  biodegradable qualities, 124–125
  BPA, 128
  composition, 124
  discarded/lost fishing nets, 125–126
  environment, 125
  EPA, 123
  phthalates, 125, 127–128
  plastic bags and jellyfish, 125, 127
  polyethylene and polystyrene, 123–124
  soda ring, 125, 127
  subtropical convergence zone, 125–126
  toxic chemicals, 125
Plumbaginaceae, 54
Polyaromatic hydrocarbons, *see* Polycyclic
    aromatic hydrocarbons (PAHs)
Polybrominated biphenyls (PBBs), 21, 84, 93–94
Polybrominated diphenyl ethers (PBDEs), 21, 84,
    93–94, 152–153, 155, 164–165
Polychlorinated biphenyls (PCBs)
  Aroclors, 87
  biological effects, 90
  biphenyl, 84–85
  breakdown of, 87–88
  characteristics, 86
  concentrations, 88–90
  congeners, 85
  coplanar/nonplanar, 86
  dioxin-like, 85
  elements, 21
  EPA, 102–103
  ortho-substitutions, 86
  persistence, 86–87
  plants and animals, 35–36
  variances, 149
Polychlorinated dibenzofurans (PCDFs),
    102–103; *see also* Furans

Polychlorinated dibenzo-p-dioxins (PCDD),
    see Dioxins
Polycyclic aromatic hydrocarbons (PAHs),
    29, 36
  BCFs, 105–106
  biological effects
    bay/fjord region, 106–107
    birds, 108
    coal tar and sealants, 108
    cytochrome P450 system, 107
    disorders, 106
    fish, 107–108
    soil particles, 105
    toxic effects, 106
  carbon and hydrogen, 101–102
  carcinogenic contaminants, 101
  chemical characteristics, 102–103
  elements, 21
  environmental concentrations, 104–105
  oil spills, 108–109
  persistence, 103–104
  priority pollutants, 102
  sources and uses, 103
Polyethylene terephthalate (PETE), 124
Polypropylene (PP), 124
Polystyrene (PS), 124
Polyvinyl chloride (PVC), 123–124
Popillia japonica, 118
POPs, *see* Persistent organic pollutants
PPCPs, *see* Pharmaceuticals and personal
    care products
Primary sex ratio, 148
Priority pollutants, 52, 102
Pseudacris regilla, 29–30
Pyrethrins, 71
Pyrethroids, 33
  birds and mammals, 74
  half-lives, 73–74
  insect toxicity, 73
  paresthesia, 74
  pyrethrum, 71–72
  respiratory failure, 74
  synthetic pyrethroids, 71, 73

Q

Quiscalus quiscula, 117

R

Rangia cuneata, 105
Regulatory agencies, 208
Resiliency, 183
Resistance, 182–183
Resource Conservation and Recovery Act
    (RCRA), 199

Risk assessment, 7, 210
 ecological risk assessment
  aquatic environments, 176–177
  contaminants, 177
  cost-benefit analysis, 178
  decision making, 178
  definition, 175
  process, 177
 environmental hazards, 175–176
 history, 176
 organisms
  chemical balance equation, 178, 180
  exposure index/exposed dose, 178–180
  hazard quotient, 180–181
  TRV, 180
 risk management, 183–185
 uncertainty, 181–182
 vulnerability analysis, 182–183
Risk management, 183–185, 210
Roundup, 75

S

Saccharomyces cerevisiae, 52
Salvelinus namaycush, 89
Scoliosis, 29–30
Seirus aurocapilla, 135
Selenomethionine, 30
Silent Spring, 5
Somateria mollissima, 108
Spartina alterniflora, 167
Statistics, 7
Stenella coeruleoalba, 89
Stockholm Convention, 190–191
Streptomyces fungi, 75
Streptopelia risoria, 93
Sturnus vulgaris, 117
Sublethal effects
 acute exposures, 4
 birds and mammals, 60
 chronic exposures, 4
 endocrine disruption/tumor production, 177
 environmental conditions, 70
 LC50 and LD50 tests, 39, 41, 59
 OCPs, 117
Superfund Site, 200
Synthetic pyrethroids, 71, 73

T

Target organism, 68
Teratogenic, 29
Terrepene carolina, 54

2,3,7,8-Tetrachlorodibenzo-p-dioxin (TCDD), 92–93
Tilapia guineensis, 151
Tissue banks, 209
Toxic equivalency factor (TEF), 92–93
Toxic equivalency quotients (TEQs), 92–93, 149
Toxicity reference value (TRV), 180, 208
Toxicology, 207
Toxic Release Inventory (TRI), 200
Toxic Substances Control Act (TSCA), 199–200, 208
Transition metals, 12, 14
Triazines, 74, 76–77
Turdus migratorius, 117
Tursiops truncates, 116

U

Uncertainty, 181–182
United Nations Environment Programme (UNEP), 189–191
United Nations Framework Convention on Climate Change (UNFCCC), 190–191
Unreasonable adverse effects, 201
Ursus maritimus, 95, 151–152
U.S. Coast Guard, 197
U.S. Forest Service (USFS), 194
U.S. Geological Survey (USGS), 195–196

V

Valence shell, 12
Vienna Convention, 191
Vulnerability analysis, 182–183

W

Web of Science™, 205
Wet deposition, 132–133
Wet weight, 23

X

Xenopus laevis, 37–38

Z

Zea mays, 165
Zinc oxide (ZnO) nanoparticles, 165